Embedded Mechatronic Systems 1

Embedded Mechatronic Systems 1

Revised and Updated 2nd Edition

Embedded Mechatronic Systems 1

Analysis of Failures, Predictive Reliability

Edited by

Abdelkhalak El Hami
Philippe Pougnet

First edition published 2015 in Great Britain and the United States by ISTE Press Ltd and Elsevier Ltd
© ISTE Press Ltd 2015.

This edition published 2019 in Great Britain and the United States by ISTE Press Ltd and Elsevier Ltd

Apart from any fair dealing for the purposes of research or private study, or criticism or review, as permitted under the Copyright, Designs and Patents Act 1988, this publication may only be reproduced, stored or transmitted, in any form or by any means, with the prior permission in writing of the publishers, or in the case of reprographic reproduction in accordance with the terms and licenses issued by the CLA. Enquiries concerning reproduction outside these terms should be sent to the publishers at the undermentioned address:

ISTE Press Ltd
27-37 St George's Road
London SW19 4EU
UK

www.iste.co.uk

Elsevier Ltd
The Boulevard, Langford Lane
Kidlington, Oxford, OX5 1GB
UK

www.elsevier.com

Notices
Knowledge and best practice in this field are constantly changing. As new research and experience broaden our understanding, changes in research methods, professional practices, or medical treatment may become necessary.

Practitioners and researchers must always rely on their own experience and knowledge in evaluating and using any information, methods, compounds, or experiments described herein. In using such information or methods they should be mindful of their own safety and the safety of others, including parties for whom they have a professional responsibility.

To the fullest extent of the law, neither the Publisher nor the authors, contributors, or editors, assume any liability for any injury and/or damage to persons or property as a matter of products liability, negligence or otherwise, or from any use or operation of any methods, products, instructions, or ideas contained in the material herein.

MATLAB® is a trademark of The MathWorks, Inc. and is used with permission. The MathWorks does not warrant the accuracy of the text or exercises in this book. This book's use or discussion of MATLAB® software or related products does not constitute endorsement or sponsorship by The MathWorks of a particular pedagogical approach or particular use of the MATLAB® software.

For information on all our publications visit our website at http://store.elsevier.com/

© ISTE Press Ltd 2019
The rights of Abdelkhalak El Hami and Philippe Pougnet to be identified as the authors of this work have been asserted by them in accordance with the Copyright, Designs and Patents Act 1988.

British Library Cataloguing-in-Publication Data
A CIP record for this book is available from the British Library
Library of Congress Cataloging in Publication Data
A catalog record for this book is available from the Library of Congress
ISBN 978-1-78548-189-5

Contents

Preface xi

Chapter 1. Reliability-based Design Optimization 1
Philippe POUGNET and Abdelkhalak EL HAMI

    1.1. Introduction 2
    1.2. Reliability-based design optimization 5
        1.2.1. Risk assessment using predictive
        reliability calculations 7
        1.2.2. Identifying elements that are critical
        for the reliability of the system 9
        1.2.3. Determination of the distribution of stresses
        leading to failures 11
        1.2.4. Determining the critical effect of stresses 14
        1.2.5. Inducing failures for failure mechanism analysis 17
        1.2.6. Failure mechanism modeling 22
        1.2.7. Design optimization 23
    1.3. Conclusion 25
    1.4. References 26

**Chapter 2. Non-destructive Characterization
by Spectroscopic Ellipsometry of Interfaces
in Mechatronic Devices** 29
Pierre Richard DAHOO, Malika KHETTAB, Jorge LINARES and Philippe POUGNET

    2.1. Introduction 30
    2.2. Relationship between the ellipsometric parameters
    and the optical characteristics of a sample 32

2.3. Rotating component or phase modulator ellipsometers 34
2.4. Relationship between ellipsometric parameters
and intensity of the detected signal. 35
2.5. Analysis of experimental data 36
2.6. The stack structural model 39
2.7. The optical model 40
2.8. Application of ellipsometry technique 43
   2.8.1. Thin layer from silver nanograins sintered
   on a copper substrate. 45
   2.8.2. Analysis of ellipsometric spectra of polymers
   on different substrates 48
   2.8.3. Analysis and comparison after stress 53
   2.8.4. Physical analysis of light and matter interaction
   in terms of band gap energy. 54
2.9. Conclusion 56
2.10. References. 57

Chapter 3. Method of Characterizing the Electromagnetic Environment in Hyperfrequency Circuits Encapsulated Within Metallic Cavities 61

Samh KHEMIRI, Abhishek RAMANUJAN, Moncef KADI and Zouheir RIAH

3.1. Introduction. 61
3.2. Theory of metallic cavities 62
   3.2.1. Definition. 62
   3.2.2. Electromagnetic field in a parallelepiped cavity 63
   3.2.3. Resonance frequencies. 63
3.3. Effect of metal cavities on the radiated emissions
of microwave circuits. 65
   3.3.1. Circuit case study: 50 Ω microstrip line. 65
3.4. Approximation of the electromagnetic field
radiated in the presence of the cavity from the
electromagnetic field radiated without cavity 71
   3.4.1. Principle of the approach 71
   3.4.2. Radiated emission model 72
   3.4.3. Results and discussion 77
3.5. Conclusion 80
3.6. References 81

Chapter 4. Metrology of Static and Dynamic Displacements and Deformations Using Full-Field Techniques 83

Ioana NISTEA and Dan BORZA

    4.1. Introduction 83
    4.2. Speckle interferometry. 86
        4.2.1. Principles of displacement field metrology by speckle
        interferometry 86
        4.2.2. Description of the speckle interferometry
        measurement setup 94
        4.2.3. Examples of static displacement field measurements 95
        4.2.4. Examples of measurements of vibration
        displacements fields. 105
        4.2.5. Examples of dynamic measurements 111
    4.3. Moiré projection 112
        4.3.1. Measurement principles of Moiré projection
        for displacement fields 113
        4.3.2. Description of the Moiré projection
        measurement setup 114
        4.3.3. Examples of displacement field metrology
        by Moiré projection 115
    4.4. Structured light projection. 116
        4.4.1. Principles of structured light projection
        for measuring surface topography 116
        4.4.2. Description of the structured light projection
        measurement setup 118
        4.4.3. Examples of surface topography measurements
        by structured light projection 118
    4.5. Conclusion. 120
    4.6. References 121

Chapter 5. Characterization of Switching Transistors Under Electrical Overvoltage Stresses 123

Patrick MARTIN, Ludovic LACHEZE, Alain KAMDEL and Philippe DESCAMPS

    5.1. Introduction 123
    5.2. Stress test over ESD/EOV electric constraints. 124
        5.2.1. Description of the TPG test equipment. 124
        5.2.2. Stresses applied to the transistor. 126
        5.2.3. Testing procedure 128
        5.2.4. TPG capabilities 129

5.3. Simulation results 129
  5.3.1. Highlighted phenomena 129
  5.3.2. Influence of parasitic phenomena 130
5.4. Experimental setup 133
  5.4.1. Measurement results and analysis of
  observed phenomena. 134
5.5. Conclusion 142
5.6. References 143

Chapter 6. Reliability of Radio Frequency Power Transistors to Electromagnetic and Thermal Stress 145

Samh KHEMIRI and Moncef KADI

6.1. Introduction. 145
6.2. The GaN technology 146
6.3. Radiated electromagnetic stress 148
  6.3.1. Presentation of the test equipment 148
  6.3.2. Results and analysis 149
6.4. RF CW continuous stress. 153
  6.4.1. Presentation of the test equipment 153
  6.4.2. Results and analysis 154
6.5. Thermal exposure 156
  6.5.1. Presentation of the test equipment 156
  6.5.2. Results and analysis 157
6.6. Combined stresses: RF CW + electromagnetic
(EM) and electric + EM 159
  6.6.1. Effect of the simultaneous application
  of EM and RF stresses. 160
  6.6.2. Effect of the simultaneous application of
  electromagnetic and continuous DC stresses. 162
6.7. Conclusion 164
6.8. References 165

Chapter 7. Internal Temperature Measurement of Electronic Components 169

Eric JOUBERT, Olivier LATRY, Pascal DHERBECOURT, Maxime FONTAINE, Christian GAUTIER, Hubert POLAERT and Philippe EUDELINE

7.1. Introduction. 169
7.2. Experimental setup 170
7.3. Measurement results. 173
  7.3.1. IR measurements 173

7.3.2. Electrical measures 176
7.3.3. Optical measurement methods. 178
7.3.4. Comparison between infrared and electrical methods 183
7.4. Conclusion. 186
7.5. References 188

Chapter 8. Reliability Prediction of Embedded Electronic Systems: the FIDES Guide. 189

Philippe POUGNET, Franck BAYLE, Hichame MAANANE and
Pierre Richard DAHOO

8.1. Introduction 189
8.2. Presentation of the FIDES guide 191
   8.2.1. Global modeling 191
   8.2.2. Generic model 191
   8.2.3. Mathematical foundations 192
   8.2.4. Justifying the use of a constant failure rate/intensity 194
   8.2.5. Assessing λ_o 195
   8.2.6. Acceleration factors 195
   8.2.7. Life profile 196
   8.2.8. Testing performed at electronic board level 199
   8.2.9. Component-level testing 201
   8.2.10. "Component family" testing 202
   8.2.11. Example of "MOSFET" power transistors 203
8.3. FIDES calculation on an automotive mechatronic system 204
   8.3.1. Goals of the FIDES calculation 206
   8.3.2. Methodology 206
   8.3.3. Life profile 207
   8.3.4. Results for the SMI board components. 212
   8.3.5. Results for the FR4 board components 212
   8.3.6. Failure rate of the DC-DC converter 213
   8.3.7. Effect of the amplitude of the thermal cycles on the lifetime 214
   8.3.8. Comparison with the results of the UTE C 80-810 standard 214
8.4. Conclusion. 215
8.5. References 215

Chapter 9. Multi-objective Optimization in Fluid–Structure Interaction 217
Rabii EL MAANI, Abdelkhalak EL HAMI and Bouchaïb RADI

    9.1. Introduction. 217
    9.2. Backtracking search algorithm 220
        9.2.1. Initialization 220
        9.2.2. Selection I 221
        9.2.3. Mutation operator. 221
        9.2.4. Crossover operator 222
        9.2.5. Selection II. 222
    9.3. Multi-objective optimization problem 223
    9.4. Proposed algorithm 224
        9.4.1. Fast non-dominated sorting 225
        9.4.2. Crowding distance 226
        9.4.3. Numerical validation 227
    9.5. Application to FSI problems 231
        9.5.1. Statement of the FSI problems 231
        9.5.2. Process of FSI optimization 235
        9.5.3. Application to the Onera M6 wing 236
    9.6. Conclusion 245
    9.7. References 246

List of Authors 251

Index 255

Preface

Electronics are increasingly used in controlled and embedded mechanical systems. This leads to new mechatronics devices that are lighter, smaller and use less energy. However, this mechatronics approach, which enables technological breakthroughs, must take into account sometimes contradictory constraints such as lead-time to market and cost savings. Consequently, implementing a mechatronic device and mastering its reliability are not always entirely synchronized processes. For instance, this is the case for systems that function in harsh environments or in operating conditions which cause failures. Indeed, when the root causes of such defects are not understood, they can be more difficult to control. This book attempts to respond to these problems. It is intended for stakeholders in the field of embedded mechatronics so that they can reduce the industrial and financial risks linked to operational defects. This book presents a method to develop mechatronics products where reliability is an ongoing process starting in the initial product design stages. It is based on understanding the failure mechanisms in mechatronic systems. These failure mechanisms are modeled to simulate the consequences and experiments are carried out to optimize the numerical approach. The simulation helps to reduce the time required to anticipate the causes of these failures. The experiments help to refine the models which represent the systems studied.

This book is the result of collaborative research activities between private (big, intermediate and small businesses) and public sector agents (universities and engineering schools). The orientations of this research were initiated by the Mechatronics Strategic Branch of the Mov'eo competitive cluster (*Domaine d'Action Stratégique*) to meet the need to have reliable mechatronic systems.

This book is aimed at engineers and researchers working in the mechatronics industry and Master's or PhD students looking to specialize in experimental investigations, experimental characterization of physical or chemical stresses, failure analysis and failure mechanism modeling to simulate the consequences of causes of failure and who want to use statistics to assess reliability. These subjects match the needs of the mechatronics industry.

This book is divided into two volumes. This volume presents the statistical approach for optimizing designs for reliability and the experimental approach for characterizing the evolution of mechatronic systems in operation. Volume 2 [ELH 19] looks at trials and multi-physical modeling of defects which show weaknesses in design and the creation of meta-models for optimizing designs.

Chapter 1 describes the reliability-driven design methodology by building on a case study. The first step in this approach is to define the reliability targets, the risks of failure due to architectural innovations or new conditions of use and then to evaluate the predictive reliability of the electronics. The objectives of the following steps are to identify the components that may fail in the life profile conditions and determine the distribution of the stresses causing these failures. In order to understand the potential failure mechanisms, experimental characterizations of the effects of mechanical, thermal or electromagnetic stresses are carried out on a few prototypes and tests are designed to provoke failures. Consecutive failure analysis helps to develop failure mechanism models. However, these multi-physical models are based on approximations and uncertainties. They have to be validated before being used to simulate the failures in the conditions of the life profile. Using statistical approaches, the multi-physical failure models can take into account the variability of the loads of the life profile as well as the variability of the manufacturing process. The design is then optimized by adjusting the architecture parameters that improve reliability. Chapter 2 describes the Spectroscopic Ellipsometry (SE) method. This method is often used in microelectronics to study semiconductors, polymer-based protective coatings, metals, or other types of meta-materials. SE is applied here to study the effect of environmental stresses on the quality of surfaces and interfaces of sintered silver materials and polymers of a mechatronic power module. A study of the effect of temperature in dry and wet environments is presented and discussed in terms of optical properties. Chapter 3 describes an approach determining emissions radiated from microwave structures found in metallic cavities. This approach is based on near-field cartographies and on a model of the emissions radiated from the open structure by a network of dipoles.

Chapter 4 presents the experimental study of the static and dynamic deformations of the components and electronic equipment, using optical techniques of coherent light based on full-field methods. The applied interferometric and non-interferometric techniques lead to complementary results in terms of temporal and spatial resolution as well as measuring sensitivity. These results have been obtained by applying the techniques of Speckle Interferometry (SI) to temporal integration, Moiré Projection (MP) and Structured Light (SL) to study the phenomena related to the thermomechanical and vibratory behaviour of the embedded electronic devices. Chapter 5 describes a method of characterizing the robustness of switching transistors relative to overvoltage electrical stresses. In this approach the phenomena of electrostatic discharge (ESD) are reproduced. Chapter 6 focuses on the study of the reliability and the robustness of radiofrequency power transistors (RF) used in power amplifier electronic boards (HPA: High Power Amplifier). These transistors are the base elements of the (Tx) transmission modules for radar applications. The effects of radiated electromagnetic waves, RF signals and thermal loads on Gallium Nitride (GaN) RF transistors are studied.

Chapter 7 presents a method for measuring temperature and micro-displacement on high frequency components used in telecommunications and radars. The simultaneity of the measurements of the temperature and expansion parameters represents the originality of this method. This approach makes it possible to calculate the thermal resistance of an electronic component and study how this resistance changes during the life of the component. Chapter 8 presents the FIDES predictive reliability handbook. FIDES approach is based on defining the life profile and provides prediction of the failure rate of mechatronic systems. FIDES is frequently updated and follows the changes of the electronic technology. FIDES is here applied to an automotive mechatronic system. Chapter 9 presents a new algorithm for optimizing retrieval search for multi-objective optimization named BSAMO. This, evolutionary algorithm (EA) solves real-valued numerical optimization problems. EAs are stochastic research algorithms widely used to solve non-linear, non-differentiable complex numerical optimization problems. In order to test its performance, this algorithm is applied to a well-known multi-objective case study. The FSI is optimized, using a partitioned coupling procedure. This method is tested on a 3D wing subjected to aerodynamic loads. The Pareto solutions obtained are presented and compared to those of the non-dominated sorting genetic algorithm II (NSGA-II). The numerical results demonstrate the efficiency of BSAMO and its ability to solve real-world multi-physics problems.

The editors would like to thank the following public bodies for supporting the AUDACE (Analysis of the Failure Causes of Embedded Mechatronic Systems) program: formerly DGCIS (*Direction générale de la compétitivité, de l'industrie et des services*) now DGE (*Direction Générale des Entreprises*), Île de France Regional Council (*Conseil régional Île de France*), Haute-Normandie Regional Council (*Conseil régional Haute-Normandie*), Basse-Normandie Regional Council (*Conseil régional Basse-Normandie*), Val d'Oise General Council (*Conseil Général du Val d'Oise*), Yvelines General

Council (*Conseil Général des Yvelines*), Essone General Council (*Conseil Général de l'Essonne*), Cergy-Pontoise Federation of Municipalities (*Communauté d'Agglomération de Cergy-Pontoise*), MOV'EO competitive cluster and Normandie AeroEspace (NAE) competitive cluster.

Abdelkhalak EL HAMI
Philippe POUGNET
July 2019

References

[ELH 19] EL HAMI A., POUGNET P. (eds), *Embedded Mechanationic Systems 2: Analyses of Failures, Modeling, Simulation and Optimization* (2nd edition), ISTE Press, London, and Elsevier, Oxford, 2019.

Council (United Nations for Iceland), Saudi General Council (Council Curatorium of Resources), Liuqa Pontoloes Federation of Manufacturing (Consumers Federation of the Cypress Council), SIOVEO cooperation States and Normadic Agricultural (SAF) cooperative statics.

Aboukbalaa El HANI,
Philippe TOUCHET
July 2019

References

[DUB, 1991] DUBAR, A.G., TOUCHET, P., (eds), Fundamental Thermodynamics, Analysis of Reaction, Reactors, Simulation and Optimisation (2nd edition) (STR Press, London and Lavoisier, Orleans 2024).

1

Reliability-based Design Optimization

In order to increase and maintain their businesses, mechatronics manufacturers develop innovative products and reduce product development costs. Economic constraints motivate them to reduce the duration of the testing phase and the number of prototypes and to develop simulations. Introducing innovations enables them to meet customers' expectations and stand out from their competitors. A poor assessment of the ability of these innovative products to function properly in the conditions of use may result in nonconformities during warranty periods that negatively impact profitability. To reduce these industrial and financial risks, reliability must be incorporated into the design process.

This chapter describes a reliability-based design methodology for embedded mechatronic systems. The first step in this approach is to define the reliability targets, the risks of failure due to architectural innovations or new conditions of use, and then to evaluate the predictive reliability of the electronics. The FIDES reliability guide, a handbook on predictive reliability based on the laws of the physics of failure, regularly updated according to field returns, provides realistic forecasts of the conditions of use. The objectives of the following steps are to identify the potentially faulty in the life profile conditions and then to determine the distribution of the constraints causing these failures. In order to understand failure mechanisms, experimental characterizations of the effects of mechanical, thermal or electromagnetic stresses are carried out on a few prototypes, and tests are designed to provoke failures. Consecutive failure analysis helps to develop multiphysics

---

Chapter written by Philippe POUGNET and Abdelkhalak EL HAMI.

failure models. These failure models are optimized and then validated by comparing model responses to thermal or vibratory solicitations with results. Developing metamodels capable of including the variability of the life profile loads and of fabrication enables reliability predictions. The design is then optimized by adjusting the architecture parameters that improve reliability.

1.1. Introduction

The mechatronics business sector is expanding rapidly due to the practice of embedding electronics inside mechanical devices, enabling manufacturers to reduce volume, mass and energy consumption, as well as production costs, which helps to gain market share. To develop new mechatronic systems, manufacturers need to face several challenges. They have to be competitive in terms of production costs and development lead time but must also ensure a high level of performance and correct functioning in increasingly stringent operational conditions for even longer lifetimes. Design and validation need to be firmly controlled.

Most frequently, the design of a mechatronic system is done by a tier one supplier in order to respect the set of requirements provided by the original equipment manufacturer (OEM). The specifications detail the requirements, the obligatory performances, use conditions, (operational and environmental, and storage and transport conditions) and the expected reliability objectives. These elements enable the tier one supplier to define the life profile which is the basis of the mechatronics system architecture [CRO 01].

After analyzing the required functions and constraints defined by the specifications [CRE 03], the designers draw up the functional modules that perform the high-level functions of the system. The performances detailed in the specifications are expressed as objectives to be met for each of the functional modules. These objectives are defined by measurable physical quantities and tolerances. In a mechatronic product design process, for each high-level functional module, the designers specify the necessary sub-functions, the constraints and the physical elements needed to perform these functions

[SUH 01]. This process of translating the functional requirements and constraints into blocks in the physical domain is repeated at the various levels of the product architecture down to the basic elements. In a mechatronic system the basic elements are the electronic components. The functions and characteristics of electronic components and the required fabrication assembly processes are also detailed.

To meet the cost and size requirements, embedded mechatronic system architectures are designed without redundancy which means that an embedded mechatronic system may fail if only one of its components is defective.

A mechatronic system is reliable if it performs the required functions under the intended conditions of use for a specified period of time. The reliability of a mechatronic system is defined by the probability of performing high-level functions of the system for a given confidence level. For example, a mechatronic system is reliable if it is able to operate without failure for 20 years with a probability of 0.9 and a confidence rate of 80%. Another reliability objective may be product lifetime as defined by the length of time that a given proportion of products operates without failure for a given confidence level. By definition, reliability can only be assessed when the product is manufactured and used. However, it is obvious that manufacturers of embedded mechatronic systems cannot wait to manufacture systems in large series to prove their reliability.

In the electronic systems industry, designers predict reliability by applying predictive reliability guides. These approaches are based on the definition of a mission profile, i.e. on the definition of the thermo-mechanical and electrical stress cycles applied on the electronic components in use and on aging models based on experience. The major advantage of applying predictive reliability guides is in identifying the electronic components that are critical to reliability because of their high predictive failure rates. However, these predictive reliability handbooks have limitations: the latest technological developments of electronic components are not always included. Some guides like the MIL HDBK 217F are still in use while their component libraries are obsolete. If field returns are insufficiently taken into account, the predicted performance will differ enormously from the operation results.

The first step in establishing the reliability of an embedded mechatronic system is to define the mission profile. All the cycles of operational and environmental stresses which are applied on the components and cause aging are listed. In some cases, it may be useful to perform physical characterizations of the effects of stresses on prototypes. These characterizations performed under conditions close to those found in application are important for identifying the components or elements that are critical for reliability. Accelerated tests are then performed. Their purpose is to evaluate the strength of critical components to the stresses of the mission profile. Failure analysis and the laws of the physics of failure are then used to understand the failure mechanism and predict the risks of failure. In the case where the required level of reliability is not achieved, the design needs to be adapted. A numerical model of the failure mechanism is then developed by the finite element method. A stochastic metamodel that approximates the actual physical response of the failure mechanism is developed. This metamodel makes it possible to adjust design parameters that optimize the probability of operating without failure and to reduce sensitivity to variability.

This chapter describes the steps of a reliable mechatronic systems design methodology. After defining the reliability targets, the life profile is analyzed. The stress cycles that will cause fatigue to the critical components are identified. Optical characterization techniques are applied to better understand the constraints applied during the conditions of use. The strength of critical components to the stresses that may occur in application is evaluated by accelerated tests. Failure analysis and the physics of failure make it possible to model the failure mechanism. The use of computer simulation based on this failure model makes it possible to take into account the variability of the conditions of fabrication and use and evaluate reliability. The design so that its probability of functioning without failure respects the requirements. This approach is illustrated by a case study.

1.2. Reliability-based design optimization

Questions about reliability can be addressed early on in the product development cycle. For instance, the following questions could be answered:

– Are the project reliability targets defined?

– Is the product life profile clearly defined?

– What are the product performances that are important and that cannot be degraded during the expected life of the product?

– What are the expected loads for a standard use (temperatures, vibrations, power cycles, thermal cycles, moisture condensation cycles, humidity, mechanical shocks, electrical stresses, etc.) that can significantly impact the performance of the product and therefore its reliability?

– Regarding the history of the product line, what in the design or the manufacturing process is new, modified or renewed?

– What are the risks for reliability (i.e. risks due to wear or aging in the conditions of use) of the new elements of the design or process?

– What is planned to eliminate the identified technical risks?

Once the system design is defined, the mission profile can be customized to each component and the potential effects of the life profile on reliability can be evaluated. A forecast reliability calculation estimates the failure rate. A review of the elements of the architecture identifies the components that are critical for reliability. The distribution of the mechanical stresses that may cause failures in the critical components can be characterized in a prototype by appropriate optical techniques of deformation measurements or by acceleration measurements by applying vibrations to the product. Temperature measurements of components activated by power cycles or on/off cycles can be performed using thermal imaging.

Reliability-based design optimization is only possible if validated models of the system and its failure mechanisms are available. A finite element model of the system can be evaluated by comparing the response of the model and measurements obtained on a prototype tested under mechanical or thermomechanical loads. An optimization method can be used to adjust design parameters and reduce the model and prototype response discrepancies and validate the model. Validation of the failure mechanism model is done by comparing the failure predictions to the failures found in tests. The design factors having a strong effect on the system failure mechanism are then identified with a screening method. A plan of numerical experiments makes it possible to develop a metamodel. This metamodel makes it possible to improve design by minimizing the effects of variability in design factors while maximizing reliability. The flow chart of this approach is shown in Figure 1.1.

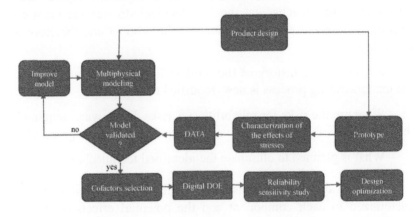

Figure 1.1. *Flow chart of the reliability-based design optimization methodology*

CASE STUDY.–

A case study based on an engine control unit (ECU) further illustrates this approach. This electronics system is located under the hood near the power train system (Figure 1.2). The ECU performs the analogical and digital electronic functions which control the thermal engine set points.

Reliability-based Design Optimization 7

Figure 1.2. *ECU located under the hood*

The ECU is an embedded mechatronic system that must operate with a reliability rate of 0.9999 for a 15-year period of use.

What is new in the architecture is a large microcontroller component with a large number of pins at a very fine pitch.

According to an analysis of the life profile, the stresses that can cause failures in the product components are the vibrations and thermal cycles accumulated during activation cycles.

To optimize the design of the ECU by reliability, a finite element model of the printed circuit board fixed by screws to its housing is developed. This model is validated by comparing the response of the model (displacements of the printed circuit during the application of temperature cycles) to the performance of a prototype under thermal loads.

1.2.1. *Risk assessment using predictive reliability calculations*

Predictive reliability calculations are used to assess the electronic system failure rate very early in its development. In these calculations, electronic component failures are assumed to be independent and random occurrences. Intrinsic failures caused by fatigue, wear or technological limitations are not considered. Figure 1.3 schematically shows the different steps of a predictive reliability calculation.

The guidelines used to assess predictive reliability in the embedded mechatronics industry are the MIL HDBK 217F standard and the UTE C 80-810 (RDF 2000) standard. These guides are accepted by most OEMs of the automotive industry. However, they have limitations. These are drawbacks – for example, they are not recent (MIL HDBK 217 standard was set up in the 1990s and the RDF 2000 standard in 2000) and they are not kept up to date. Their reliability predictions are not closely correlated with recent field return data [BNA 05]. In order to make better predictions available, Chapter 2 of this volume outlines the FIDES standard [GRO 09]. FIDES, founded in 2009, has proven to be useful in the fields of defense and aeronautics. It is updated regularly and so continues to evolve. Figure 1.3 demonstrates how to use FIDES standard.

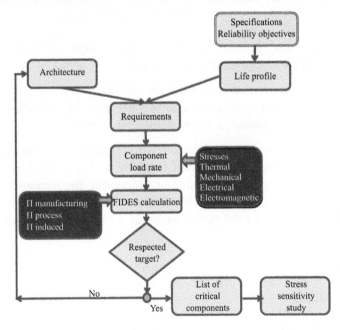

Figure 1.3. *The different stages of a FIDES forecast reliability calculation*

CASE STUDY.–

Figure 1.4 displays the failure rates expressed in failure in time (FIT) (1 FIT equals one failure per 10^9 h) of the components of the ECU having a high predictive failure rate. These are, by decreasing order of importance:

– the microcontroller;

– an application-specific integrated circuit (ASIC);

– a connector;

– a power transistor (MOSFET);

– etc.

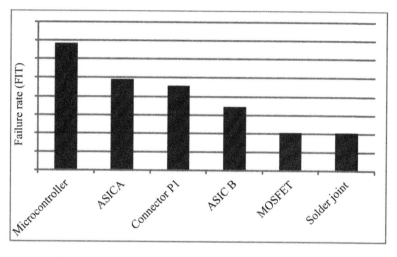

Figure 1.4. *Predictive failure rates (1 FIT equals one failure per 10^{-9} h) of the ECU critical components*

1.2.2. *Identifying elements that are critical for the reliability of the system*

The elements that are critical for reliability are components that under the conditions of use are likely to have intrinsic failures (wear, aging, fatigue and corrosion). These are either innovative components lacking sufficient information as to their robustness, or components that have been known by field returns as having a high failure rate.

Design elements that are critical for reliability are identified by checking the Bill of Materials (BOM) for the system. These critical components are listed in a document called the reliability matrix. This document describes, for each critical component, the expected modes and mechanisms of failure as well as the stresses that have a significant effect on these failures. The tests that are necessary to verify the robustness of the design elements with regard to the stresses causing failures are also documented. These tests constitute the reliability validation plan. The procedure for identifying elements critical for reliability is as follows:

– identify the expected failure modes of the critical part in operation (this means describing failure symptoms such as open circuit, short circuit, leakage current, parameter drift, degradation, etc.);

– describe the physical appearance of the defect (what is observed on the part when failure is analyzed: fissure, wear, oxidation and contamination) to clarify the way in which a part has failed;

– identify the expected failure mechanism (this means the physical and chemical causes of failure: fatigue, breakage, creep, deformation, wear, degradation due to exposure to solar or electromagnetic radiation, chemical corrosion, electromigration, electrostatic discharge, electrical overstress, insulation breakage, etc.);

– identify the stresses which trigger the failure mechanism (temperature level, number of thermal cycles, duration of exposure to ultraviolet (UV) radiation or moisture, number of dew cycles, current density amplitude, electric potential level, number of friction cycles, number of mechanical shocks, strength and duration of vibrations, number of pressure cycles, duration of exposure to corrosive agents, etc.);

– define the test which will lead to this failure mechanism;

– outline the test protocol (equipment and procedures) to track functioning failures;

– report whether the acceleration factor of the failure mechanism is known or not.

CASE STUDY.–

Table 1.1 displays the ECU reliability matrix.

The critical element is the microcontroller. This component, housed in a plastic quad flat package (PQFP), has wide dimensions (40 mm× 28 mm) and a fine lead pitch (0.5 mm). It is the first time this component is included in the BOM of an ECU. The potential failure risk for this component is the breakage of the solder joints connecting its leads to the printed circuit board (PCB). This failure mechanism may be due to fatigue caused by the accumulation od thermomechanical stresses. The stresses provoking this failure mechanism are the accumulation of temperature variation cycles.

Risk	1
Part at risk	Microcontroller
Expected Failure Mode	Open circuit
Defect Signature	Broken solder joint
Stress triggering failure mechanism	Thermo-mechanical stress
Expected failure mechanism	Breakage of solder joint due to fatigue
Stress governing failure mechanism	Temperature variations
How to test for this failure mode	-40/125°C1h/1h thermal shocks
Device to be tested	Printed circuit board in its housing

Table 1.1. *ECU reliability matrix*

1.2.3. Determination of the distribution of stresses leading to failures

The reliability matrix lists for each expected failure mode the stress or the combination of stresses which are likely to have a strong effect on the failure. To determine the distribution of the stresses that will be required to study the robustness of the critical components, the operational conditions (standard and worst case uses) and environmental conditions (thermal and mechanical stresses applied globally to the vehicle (climate, vibrations, etc.) and locally to the subsystem) are analyzed. Thermal or vibratory measurements may also be performed.

CASE STUDY.–

The vehicle has an expected life duration of 15 to 20 years depending on the type of vehicle and the manufacturer. The total distance traveled by the vehicle during its lifetime is 200,000 km. The number of hours in operation is approximately 9,000 h. The reliability requirements of the vehicle may be expressed as a maximum field return rate during the warranty period, a maximum proportion of defective devices per year, a percentage of cumulative failure after 10 or 15 years, or a warranty period lasting 15 years.

The ECU is a subsystem of the vehicle. Its architecture is designed to perform specified functions. The ECU consists of three subassemblies: an electronic board, a mechanical housing providing protection and sealing and a connector. Figure 1.5 shows schematically how the reliability requirements of the ECU fit into those of the vehicle and how the reliability of the ECU depends on the reliability of its components and their interconnections. It also displays the factors which at various levels have an impact on the expected failure mechanism.

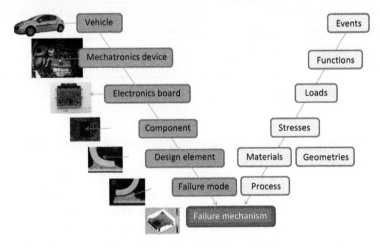

Figure 1.5. *Diagram describing ECU reliability at various architecture levels*

Each time the vehicle is used, thermal loads are applied on the ECU board. During a duty cycle, the temperature of the ECU components varies from an initial temperature to a final temperature.

The initial temperature is the temperature of the vehicle when it is parked (for example 15°C). The final temperature is the equilibrium temperature reached by the ECU board when the vehicle is in the operating mode (for example 75°C). For each standard duty cycle, the variation in temperature of the components is on average 60°C. These temperature cycle loads create thermomechanical stresses on the solder joints of the ECU components. These thermomechanical stress cycles wear the solder joints out, producing cracks that cause connection failures. The parameters to optimize the reliability of solder joints depend on the manufacturing process (choice of the solder paste, controlling the conditions under which the solder paste is deposited and the conditions of thermal reflow) as well as design variables such as the dimensions of the printed circuit solder lands.

To determine the distribution of the stresses causing the expected solder joint failure mode, the magnitude and number of temperature cycles applied on the ECU components during a standard vehicle lifetime are determined. The amplitude of the thermal cycles is set from temperature measurements on instrumented vehicles and climatic data statistics. Figure 1.6 shows the probability density of the temperature board near the components of the ECU. The number of cycles is extracted from the ECU mission profile.

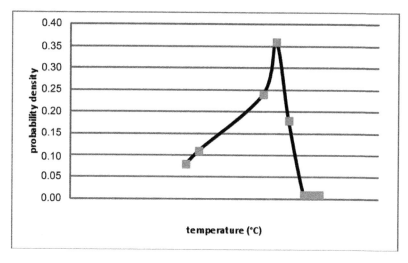

Figure 1.6. *Density probability of temperature near the components of the ECU*

1.2.4. *Determining the critical effect of stresses*

Characterization of the effect of stresses on the structural elements is used to understand the failure mechanisms of mechatronic devices. Thermal stresses and vibrations create displacements which are either in the plane or out of the plane of the structural elements. Optical methods are particularly efficient to characterize these displacements because they are contactless and not intrusive. These methods are able to consider the whole device while having a very good resolution.

CASE STUDY.–

Stresses applied on the ECU have two origins: temperature and vibrations. The power train system and the vehicle movement are sources of oscillations which deform the ECU elements. These cyclic deformations risk damaging solder joints of electronic components, condensers, terminals or connectors. Displacements caused by cyclic vibrations should be restrained. To limit the PCB displacements, the printed circuit is screwed into the casing. This increases the stiffness of the circuit board and its resonant frequency. The ECU microcontroller, which has 256 input/output pins at a 0.5 mm pitch, is placed in the middle of the PCB to facilitate routing.

To evaluate the potential risks on reliability of this component location and orientation, the out-of-plane displacements are characterized when the assembled printed circuit board is exposed to temperature variations. These displacements are studied by speckle interferometry which is a full field high-resolution holographic method (Chapter 4). Figure 1.7 shows the full-field images obtained during the cooling of the board. A break in the alignment of the fringes can be observed on the microcontroller. This break reveals a change in the out-of-plane displacements.

Figure 1.8 shows the deformations of the PCB. The position of the five screws is visible (one in the center and one at each corner). The position of the microcontroller corresponds to the dotted line. When the board is exposed to temperature variations, the printed circuit flexes but the central screw locally prevents any displacements. This leads to the PCB bending. The break which is observed in the fringes

(Figure 1.7) shows that the body of the microcontroller does not move. The leads of the component have to adapt to the local change of curvature. Strong mechanical stresses are applied to the solder joints which attach the leads to the copper prints on the circuit (Figure 1.9). Such cyclic stresses can set up traction forces in the pins and shear forces in the joints leading to fissuring of the solder joint and loosening of the pin. This causes an open circuit.

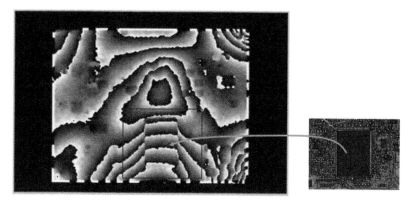

Figure 1.7. *ECU spatial distribution of fringes revealing out-of-plane displacements*

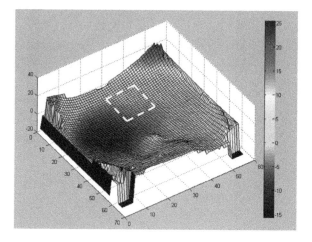

Figure 1.8. *Out-of-plane displacements (dotted line: microcontroller position). For a color version of this figure, see www.iste.co.uk/elhami/embedded1b.zip*

16   Embedded Mechatronic Systems 1

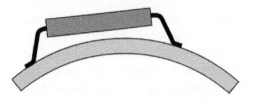

Figure 1.9. *Diagram representing how deformations affect the lead solder joints*

A second optical technique which characterizes the effect of stresses is three-dimensional (3D) laser vibrometry (Figure 1.10). In this technique, the structural elements of the mechatronic devices are submitted to vibrations by an external device and their three-dimensional movements are measured using Doppler laser beams. The dynamic response of the device is characterized using meshing. This optical technique is applied to the ECU to analyze modal deformations of the board and the components. Different integration levels are studied: bare PCB, board equipped with its components and board incased in its housing.

Figure 1.10. *Characterization of ECU by 3D laser doppler vibrometry*

A third optical technique used to measure the mechanical deformations caused by variations in temperature is the correlation of digital image (CDI). This technique is used to measure the coefficient of thermal expansion of the two critical elements of the ECU: the microcontroller and the PCB. CDI consists of seizing successive images of the deformations of the part when it is submitted to thermal changes. This technique is sensitive. It provides useful information for failure mechanism modeling.

The displacement graphs obtained show that the movements of the different points (printed circuit and component in horizontal and vertical directions) do not vary linearly along the device. Certain expansions are not constant. This means that the global value of the coefficient of thermal expansion obtained on the scale of the device varies from one point to another according to the materials present in the respective zones within the structures tested. The overall calculated values, on average 35×10^{-6} K^{-1} for the printed circuit and 5.1×10^{-6} K^{-1} for the microcontroller, vary according to the difference between the initial and final temperatures (Figure 1.11).

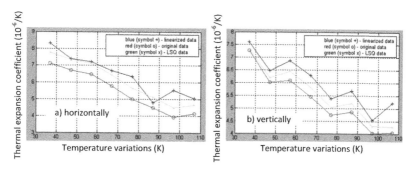

Figure 1.11. *ECU thermal expansion coefficient relative to temperature variation: a) horizontal direction and b) vertical direction. For a color version of this figure, see www.iste.co.uk/elhami/embedded1b.zip*

1.2.5. *Inducing failures for failure mechanism analysis*

In mechatronics industry design, reliability of a system is usually tested using a pass/fail endurance test, or life test. This test is typically done on a sample of parts. The stresses applied in these tests are similar to those occurring in use functioning. An acceleration factor is used to adapt the test duration to the system expected life in real use.

If none of the tested parts fail, design is evaluated as sufficiently reliable. If one part fails, design is considered unreliable.

CASE STUDY.–

The design reliability test is done on a sample of six parts. The test conditions imitate a harsh functioning set point in which the microcontroller initiates commands and tracks the result. Temperature cycles vary between 95 and –30°C (Figure 1.12). The dwell time at 95°C is 55 min. At –30°C, it is 30 min. The speed of temperature change is 5°C per minute. The power voltage varies between 0 and the power voltage of a vehicle battery. A cycle lasts 135 min.

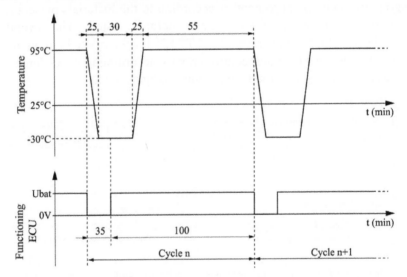

Figure 1.12. *ECU endurance test: (top) duration (in minutes) and amplitude of thermal cycles temperature (°C); (bottom) applied voltage (volts) in a cycle*

If none of the six parts is faulty after 1,000 successive cycles (the equivalent of 1,250 h test time), then the product can be considered reliable in application. This approach is very limited; it is not adapted to innovative products for which there exists no field return. It does not provide information as to how much confidence can be placed in the predictions or as to the probability of failure. Moreover, if this test is successful, design is validated as reliable even though its failure mechanisms remain unknown. A more valid approach is to define tests that induce failures. Combining failure analysis with the physics and chemistry of failure helps to understand failure mechanisms. Statistical calculations done using the data from the defective parts clearly evaluate the rate of survival under the test conditions and predict the rate of survival in the life profile conditions.

Two types of tests are used to induce failures and understand failure mechanisms:

– highly accelerated tests;

– accelerated tests.

1.2.5.1. Highly accelerated testing

Highly accelerated tests (HALT or HAT) are step-stress tests for limits. These tests are carried out in the early phases of design development to discover design weaknesses, compare architectures or technologies and check that innovative solutions are effective. These tests lead to lower field returns.

CASE STUDY.–

Table 1.2 displays highly accelerated test results of the ECU and compares these with results from previous ECU designs (types A, B and C).

1.2.5.2. Accelerated testing

In accelerated tests, stress levels are close to those found in the operational conditions. Accelerated tests are used to understand the failure mechanisms that are likely to occur in use conditions and to predict the probability density of failure in the life profile conditions. To shorten the testing time, the levels of the stresses having a significant effect on the failure mechanisms are slightly higher than the levels found in use conditions. This requires known acceleration factors.

CASE STUDY.–

Thermal shocks are used in the accelerated tests on the ECU. The temperature cycles include rapid changes between –40 and 125°C. Temperature dwell time at each extreme is 1 h. Tests for failure are performed periodically. Figure 1.13 displays the cumulated failure distribution and its 95% confidence interval versus the accumulated number of temperature shocks.

1.2.5.3. *Failure analysis*

Failure analysis is used to understand the failure mechanisms.

Highly accelerated testing	ECU	ECU type A	ECU type B	ECU type C
Cold temperature limit	–100°C	–100°C	–70°C	–65°C
High temperature limit	150°C	140°C	150°C	150°C
Search for weaknesses from five consecutive high ramp thermal cycles	No defect	No defect	No defect	No defect
Search for weaknesses from random vibrations	No defect	30 G_{RMS}	30 G_{RMS}	No defect
Search for weaknesses from stresses combining temperature and vibrations	No defect	30 G_{RMS}	30 G_{RMS}	No defect

Table 1.2. *Comparison of the results of several engine control units*

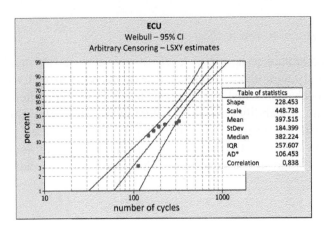

Figure 1.13. *Cumulated failures versus cumulated number of thermal cycles*

CASE STUDY.–

Failures observed in accelerated testing are disconnected leads of the microcontroller component (Figure 1.14). Failure analysis confirms that the breakage occurs at the intermetallic layer of the solder joint (Figure 1.15). This failure mechanism is not due to a defect in the assembly process but due to thermomechanical stress fatigue.

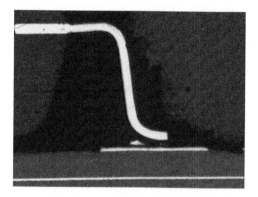

Figure 1.14. *Observed failures*

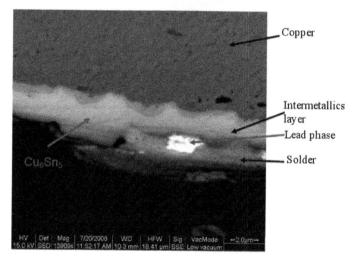

Figure 1.15. *ECU case study: scanning electron microscope observation of a micro-section of a failed solder joint*

1.2.6. *Failure mechanism modeling*

The finite element method is used to model failure mechanisms. This approach [AOU 12] efficiently calculates stresses and thermomechanical strains. The calculations of the applied stresses are used as an input for physical models of the failure mechanism. Often, based on the concept of accumulation of damage and on nonlinear material properties, these physical and stochastic models provide the density of probability of failure relative to thermomechanical stresses.

CASE STUDY.–

The failure mechanism of the solder joints of the ECU is modeled using the finite element method. The model (Figure 1.16) is composed of the PCB (composite material with different copper layers, pads connecting the microcontroller leads), the microcontroller component with its leads and the position of the screws attaching the PCB in its metallic casing. This model calculates the mechanical stresses and the strains in the solder joints when temperature varies.

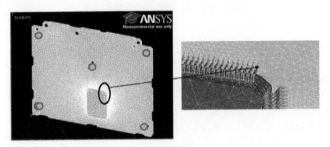

Figure 1.16. *ECU finite element model. For a color version of this figure, see www.iste.co.uk/elhami/embedded1b.zip*

The solder joint which undergoes the greatest thermomechanical stress is the joint positioned on the component diagonal in the central area of the board. Strains and stresses in this joint are calculated when five consecutive cycles of thermal loads are applied. The physical model of the failure mechanism takes into account the viscoplasticity of the solder material and creep. The calculation provides the amplitude of the plastic strain average per cycle. According to the

Coffin–Manson fatigue model, the number of cycles N_f before breakage is a function of the plastic strain ε_p as in the following equation:

$$\frac{\varepsilon_p}{2} = a\left(2N_f\right)^e \qquad [1.1]$$

In this equation, a is the fatigue ductility coefficient and e the fatigue ductility exponent of the solder material [AOU 12].

The finite element model is based on multiple hypotheses which risk introducing errors. To validate the model, the parameters are treated as uncertainties, or susceptible to variation, during the manufacturing process [MAK 12]. These parameters are described as random variables having statistical distributions. A digital design of experiments (DOE) is drawn up and the resonant frequencies of each virtual experiment are calculated. These resonant frequencies characterize the theoretical vibration behavior of the ECU. A sensitivity study identifies the parameters that significantly impact the difference between theoretical and experimental results. When the first 10 resonant frequencies cover a difference of less than 3%, the model is validated.

Thermomechanical stresses applied to the solder joints of the ECU are simulated using the data in the life profile. The failure mechanism model is then used to obtain the statistical distribution of the number of cycles to failure in the use conditions.

1.2.7. *Design optimization*

As it is not realistic to use the finite element model for Monte-Carlo-type simulations, a metamodel is developed. This metamodel replaces and rapidly calculates the response of the finite element model.

CASE STUDY.–

The metamodel of the failure mechanism of the microcontroller pins solder is used to improve the design. The various steps involved in developing this metamodel include selecting the appropriate design factors (cofactors), setting up DOE and calculating the statistical distribution of the response (number of thermal cycles leading to breakage). A sensitivity study enables the metamodel to include only the cofactors which have a significant effect on the variability of the response and also to reduce the Monte Carlo-type statistical calculation time.

Factors that can improve the response of the metamodel are design parameters, or manufacturing process parameters, such as:

– the position of the middle screw (design factor);

– the thickness of the solder (manufacturing process factor);

– the solder material composition (manufacturing process factor);

– the dimensions of the copper pads for the attachment of the microcontroller leads on the PCB (design factor).

Simulations show that possible changes in central screw location do not improve reliability. The increase in solder thickness can only be limited. The composition of the solder material has a significant effect on reliability [AOU 14a]. This option to change the solder material is ruled out due to its impact on manufacturing costs. The design factor that has the greatest impact on reliability and that has a limited impact on manufacturing costs is the dimensions of the microcontroller pin solder lands (Figure 1.17) [AOU 14b]. Indeed, by optimizing the design parameters L_{sr} and L_{sl} of the length of the copper pads and the thickness of the solder (depending on process parameters), reliability is significantly improved.

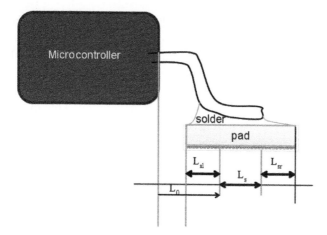

Figure 1.17. *Parameters impacting reliability: solder thickness and parameters L_{sr} and L_{sl}*

1.3. Conclusion

In both the automobile and aeronautical industries, mechatronic manufacturers are challenged to develop increasingly complex architectures within ever shorter deadlines. These systems operate in extremely difficult conditions and are required to function without any breakdown during their lifetime.

This chapter presents the various steps in reliability-based design optimization adapted to mechatronic systems. An example of this approach is the case study of an ECU. To comply with the performances required, this system uses a wide dimension fine lead pitch PQFP microcontroller. The ECU operational conditions are harsh. Predictive reliability calculations demonstrate that stresses due to cyclical temperature variations can damage the solder joints of the microcontroller leading to system breakdown. Characterizations of out-of-plane displacements set up by temperature variations are done by a speckle interferometry technique on an ECU prototype. These experiments help us to understand the failure mechanism of the solder joint of the microcontroller. Accelerated life tests using thermal shocks confirm that the solder joints are the weakest elements of the design.

A finite element model of the ECU is developed to improve the design reliability. This model is optimized and then validated by readjusting to the experimental results. Coupled with physical laws of the viscoplastic behavior of solder materials, this model simulates the failure mechanism of the solder joints of the microcontroller. A metamodel of the failure mechanism is developed. This metamodel is probabilistic and takes into account the uncertainties of the adjustment factors of the design. It makes it possible to simulate the effects on the reliability of the parameters of the design, which are in this case the dimensions of the copper pads attaching the microcontroller leads. The results of the simulations show that reliability can be improved by optimizing the length of the solder lands on the printed circuit.

This case study illustrates that reliability-based design can develop mechatronic systems which are capable of attaining the high levels of performance required under extreme environmental conditions for long periods.

1.4. References

[AOU 12] AOUES Y., MAKHLOUFI A., EL HAMI A. et al., "Probabilistic assessment of thermal fatigue of solder joints in mechatronic packaging", *Proceedings of the 1st International Symposium on Uncertainty Quantification and Stochastic Modeling*, Maresias, São Sebastião, SP, Brazil, 2012.

[AOU 14a] AOUES Y., MAKHLOUFI A., EL HAMI A. et al., "Robustness study of solder joints of different compositions by using stochastic finite element modeling", *8th International Conference on Integrated Power Electronics Systems*, Nuremberg, pp. 206–212, 2014.

[AOU 14b] AOUES Y., MAKHLOUFI A., POUGNET P. et al., "Reliability-based design optimization of the solder joint of mechatronic systems under cyclic thermal loading", *Uncertainties 2014 Conference,* Rouen, France, 23–27 June 2014.

[BNA 05] BNAE RG, Aero 0029 – Guide pour la définition et la conduite d'essais aggravés, Technical standard, Issy-les-Moulineaux, 2005.

[CRE 03] CREVELING C.M., SLUTSKY J.L., ANTIS D., *Design for Six Sigma in Technology and Product Development*, Prentice Hall, Upper Saddle River, 2003.

[CRO 01] CROWE D., FEINBERG A., *Design for Reliability*, CRC Press, New York, 2001.

[FID 09] FIDES GROUP, Reliability Methodology for Electronic Systems, Issue A, Internal report, Paris, 2009.

[HOB 05] HOBBS G.K., *HALT and HASS Accelerated Reliability Engineering*, Hobbs Engineering Corporation, Westminster, 2005.

[KEC 03] KECECIOGLU D., *Robust Engineering Design-by-Reliability with Emphasis on Mechanical Components Structural Reliability*, vol. 1, DEStech Publications, Lancaster, 2003.

[MAK 12] MAKHLOUFI A., AOUES Y., NISTEA I. et al., "Numerical and experimental investigation of the dynamic behavior of electronic systems", *1st International Symposium on Uncertainty Quantification and Stochastic Modeling*, Maresias, São Sebastião, Brazil, 2012.

[MCL 09] MCLEAN H.W., *HALT, HASS, and HASA Explained Accelerated Reliability Techniques*, American Society for Quality, 2009.

[NEL 82] NELSON W., *Applied Life Data Analysis*, John Wiley & Sons, New York, 1982.

[PRE 07] PREUSSGER A. (ed.), *Handbook on Robustness Validation of Semiconductor Devices in Automotive Application*, ZVEI Electronic Components and Systems, Frankfurt, 2007.

[SUH 01] SUH N.P., *Axiomatic Design Advances and Applications*, Oxford University Press, New York, 2001.

2

Non-destructive Characterization by Spectroscopic Ellipsometry of Interfaces in Mechatronic Devices

Failures often originate at the interfaces in the packaging of mechatronic devices because of the differences in the thermo-mechanical properties of the materials in contact. To enhance the reliability of these devices, it is necessary to have non-destructive analysis techniques such as spectroscopic ellipsometry (SE) to probe the quality of the surfaces and interfaces when the environmental constraints vary. In the field of characterization spectroscopy, SE has become indispensable in microelectronics, as well as in the study of semiconductors, protective coatings based on polymers, metals or other types of metamaterials. It is used to characterize thin films, mono- or multi-layers and bulk materials from a structural and optical point of view by probing transitions, whether electronic, vibrational or rotational. In the ultraviolet (UV)-visible and near- to mid-infrared (IR) ranges, which correspond, respectively, to electronic and vibrational absorption, SE reveals the composition, structure (amorphous or crystalline), porosity and morphology (density, roughness, etc.) of materials as a function of the light wavelength. In order to be used, SE parameters generally require an inverse method based on the simplex, the Levenberg–Marquardt or the Broyden–Fletcher–Goldfarb–Shanno algorithm to determine the dielectric function or complex optical constants. This chapter describes the SE technique and illustrates its application with two examples of characterization, that of sintered silver and polymers present in a mechatronic device. A study of the effects of temperature in dry and wet conditions is also presented and discussed in terms of optical properties.

Chapter written by Pierre Richard DAHOO, Malika KHETTAB, Jorge LINARES and Philippe POUGNET.

2.1. Introduction

In the field of optics and spectroscopy, several techniques are now routinely used for non-destructive characterization of materials, *in situ* during development or *ex situ* from a sample. We can cite, for example, interferometry, reflectometry, Raman spectroscopy, photoluminescence or spectroscopic ellipsometry (SE) which have become indispensable in the field of semiconductors, microelectronics, protective polymer coatings or metals and, more recently, nanotechnology.

Ellipsometry is used to study the interaction of polarized light with matter, usually in reflection mode. When polarized light is incident on the surface of a reflecting material, a change in its polarization state, which is the temporal evolution of the electric field vector associated with the light wave, is observed after reflection. If the polarization of the incident wave is rectilinear, for example (the electric field vector thus moves in a straight line), the reflected wave is in general elliptically polarized, which means its electric field vector describes an ellipse (Figure 2.1).

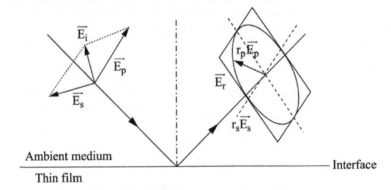

Figure 2.1. *Elliptical polarization of a linear polarized light wave after its reflection*

Ellipsometry consists of measuring this ellipticity by means of parameter $\rho = \frac{|r_p|}{|r_s|} \exp(i\Delta)$, a complex number which is the ratio of the reflection coefficients for p and s polarizations (Figure 2.1: E_p and

E_s). The theoretical approach came from Drude [DRU 87] in the late 19th Century, but the term "ellipsometry" first appeared in 1945 when Rothen [ROT 45] used this technique to distinguish methods of measurement by polarimetry.

Ellipsometry is sensitive to surfaces and interfaces, and is non-destructive [AZZ 77, DRE 82, BOC 93, PIC 95, VED 98, TOM 99, OHL 00, DAH 03, FUJ 07]. It can be applied in *ex situ* mode for characterizing materials or *in situ* mode to monitor in real time the growth of a thin layer or the effect of an external constraint (temperature, pressure, electric field, magnetic field, adsorption gas, etc.) on the evolution of the physical properties of a material (changes in composition or electronic structure, phase transition, etc.). It is used in areas as different as physics, chemistry, materials sciences, photography, art, biology, optical engineering, electronics, mechanical engineering, metallurgy, biomedicine, nanotechnology or biotechnology.

To introduce the SE technique, this chapter describes the relationship between the ellipsometric parameter ρ and the optical parameters of a material and shows how to obtain these optical characteristics experimentally. Then the different types of commercial ellipsometer are presented, and the operation of a rotating analyzer ellipsometer is described in a simplified manner to show the link between $\rho = \frac{|r_p|}{|r_s|} \exp(i\Delta)$, the fundamental relationship of ellipsometry, and the detection parameters. The method used to analyze the experimental data through an inversion procedure using a nonlinear method from ρ is then explained, and finally two examples of materials, a metallic material consisting of nano-silver grains on a copper substrate and a polymer-based material on various substrates, are analyzed by SE. The results are given and discussed for samples in the form of thin layers on substrates as a function of the temperature in wet and dry environments in terms of optical properties.

2.2. Relationship between the ellipsometric parameters and the optical characteristics of a sample

In general, the polarization state of a wave can be split into two types, one of p-type (TM wave, transverse magnetic) parallel to the plane of incidence and the other of s-type (TE wave, transverse electric) perpendicular to the plane of incidence. The amplitude of the electric fields of the reflected and transmitted waves with respect to the incident wave can be calculated (Fresnel relations), and the amplitude reflection coefficients can then be deduced as given by:

$$\text{p-type wave:} \; r_p = \frac{E_{rp}}{E_{ri}} = \frac{\tilde{n}_1 \cos\theta_i - \tilde{n}_0 \cos\theta_r}{\tilde{n}_1 \cos\theta_i + \tilde{n}_0 \cos\theta_r} \quad [2.1]$$

$$\text{s-type wave:} \; r_s = \frac{E_{rs}}{E_{ri}} = \frac{\tilde{n}_0 \cos\theta_i - \tilde{n}_1 \cos\theta_r}{\tilde{n}_0 \cos\theta_i + \tilde{n}_1 \cos\theta_r} \quad [2.2]$$

where $\tilde{n}_0$ and $\tilde{n}_1$ are, respectively, the complex optical constants of the environment and the material and θ_i and θ_r are, respectively, the angles of incidence and refraction. Generally, a material is characterized by a complex index, $\tilde{n} = n - ik$, where the real part n is related to the dispersion of light and the imaginary part k is related to the absorption of light. If the reflection coefficients of p- and s-waves are expressed in the form of a complex number comprising a magnitude term and a term of phase delay subsequent to reflection, then:

$$r_p = |r_p| \exp(i\varphi_{rp}) \quad [2.3]$$

$$r_s = |r_s| \exp(i\varphi_{rs}) \quad [2.4]$$

where the modules represent the change in amplitude of each wave type and φ_{rp} and φ_{rs} represent their phase delay. Thus, it can be shown that ρ can be written as follows:

$$\rho = \frac{r_p}{r_s} = \frac{|r_p|\exp(i\varphi_{rp})}{|r_s|\exp(i\varphi_{rs})} = \frac{|r_p|}{|r_s|}\exp(i\Delta) \quad [2.5]$$

Only the real part of ρ that is physically meaningful is taken into account in ellipsometry, and so the above formula can be expressed as: $\rho = \tan\Psi \cos\Delta$, where $\tan\Psi$ characterizes the attenuation ratio of a TM (p) wave and a TE (s) wave after reflection on the sample, and Δ is the relative phase shift. The variation range of Ψ is from 0 to 90° and that of Δ is from –180 to 180°.

In the case of a substrate:

$$\tilde{n}_1 = \tilde{n}_0 \sqrt{1 - \frac{4\rho \sin^2 \theta_i}{(1+\rho)^2}} \quad [2.6]$$

If the material is a thin layer (medium 1) on a substrate (Figure 2.2), the parameter ρ is calculated by considering only the waves that are reflected from interfaces, as indicated by the arrows in Figure 2.2.

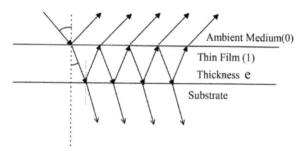

Figure 2.2. *Reflections at the interfaces in the case of a thin layer on a substrate*

By considering the ambient environment and the substrate as being homogeneous and isotropic dielectrics of respective optical indices, $\tilde{n}_0$ and $\tilde{n}_2$, and by taking into account the type of s or p polarization of the incident wave, then:

$$r_s = \frac{r_{01s} + r_{12s} e^{-i2\delta}}{\left(1 + r_{01s} r_{12s} e^{-i2\delta}\right)} \quad [2.7]$$

$$r_p = \frac{r_{01p} + r_{12p}e^{-i2\delta}}{\left(1 + r_{01p}r_{12p}e^{-i2\delta}\right)} \qquad [2.8]$$

with $\delta = (2\pi/\lambda)\, n_1 e\, \cos(\theta_1)$, where λ is the the wavelength of the incident light, n_1 is the real part of $\tilde{n}_1$ (the refractive index of medium 1), and θ_1 is the refraction angle in this medium. Note that angles θ_0, θ_1 and θ_2 are deduced from each other by Descartes–Snell's law. Finally, using $r_p = |r_p|e^{i\Delta_{r_p}}$ and $r_s = |r_s|e^{i\Delta_{r_s}}$, the ellipsometric parameter ρ of the sample can be calculated.

2.3. Rotating component or phase modulator ellipsometers

Whatever the technique used to determine the polarization state of light after its reflection on a sample, an ellipsometer consists of the same optical devices, which comprise two arms and a sample holder as shown in Figure 2.3. One arm includes a light source and a set of optical devices to transform the light wave into a known polarization state (P). The second arm comprises an analyzer (A), used to determine the state of polarization of the light wave after its reflection on the sample (E) to be analyzed.

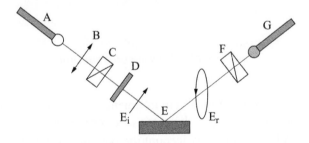

Figure 2.3. *PCSA ellipsometer in reflection mode. A: fiber source, B: lens, C: polarizer, D: compensator, E_i: incident field, E: sample, E_r: reflected field, F: rotating analyzer and G: fiber to detector*

With monochromatic light, to cancel the effect of the sample on the polarized state of the reflected light, a quarter-wave plate is used as the compensator (C) either on the side of the polarizer (PCSA mode) or on the side of the analyzer (PSCA mode). Measurements are performed either manually, by the extinction method by which one of the elements (P, C or A: rotating polarizer compensator and analyzer) is fixed so that the other two can be rotated in order to reduce to zero the light intensity after A, or automatically with a suitable detection system, by rotation of P, C or A or by phase modulation by fixing P and A and inserting a modulator, typically a quartz plate or rod subjected to an alternating electric field, after the polarizer. Whereas the first ellipsometers were monochromatic and/or had a variable angle, those commercially available today are mainly designed for SE, usually in the spectral range from near-UV to near-infrared (IR) (200–1,000 nm). They are automated (rotating component or phase modulator) and the light signal at the exit of the apparatus, after passing through a diffraction grating, is sent to a diode array or photomultipliers. The compensator in ellipsometers with rotating optical components is a device which must act as a quarter-wave plate for the entire spectral range of measurement. The uncertainties of this operating mode are integrated into a correction mode by the driving software of the instrument during data acquisition. It requires a calibration procedure of the effects of the different optical components before each measurement. The Fourier transform of the detected light signal leads, for each wavelength, to the two ellipsometric parameters, $\tan(\psi)$ or ψ and $\cos(\Delta)$ or Δ.

2.4. Relationship between ellipsometric parameters and intensity of the detected signal

Starting from the system shown in Figure 2.3 (rotating analyzer mode), the ellipsometric reflection method consists of measuring, at a fixed angle of incidence, the intensity I_R as a function of the rotation angle of the analyzer A. Denoting the electric field vector of the incident light at the entrance of the apparatus by E_0 and the electric field vector at the exit of the device after the analyzer by E_R, the latter can be determined by the following matrix relation:

$$\vec{E}_R = P_p(A)ECP_p(45°)\vec{E}_0 \qquad [2.9]$$

where each optical component is modeled by its 2 × 2 transformation matrix (Jones matrix), with E the matrix of the sample that interacts with light; C the matrix of the compensator; $P_p(45°)$ the polarizer P, whose optical axis is oriented at 45° relative to the axes of the reference system (p,s); $P_p(A)$ the polarizer P that rotates with its optical axis, making an angle that varies with respect to the axis p of the reference frame (p,s). Taking into account the rotation of the analyzer, the detected light energy or intensity I_R that is proportional to the square of the electric field E_R ($I_R = E_R E_R^*$) is a sinusoidal function of the azimuth of the analyzer of period 180°. Thus, the detected light flux varies periodically with time at twice the frequency (2f) of the rotating analyzer. The relationship between the parameters determined by the Fourier transform of the detected light signal and the ellipsometric parameters tan(ψ) or ψ and cos(Δ) or Δ can be established from:

$$I_R = \alpha + \beta\cos 2A + \chi\sin 2A + \delta\cos 4A + \varepsilon\sin 4A + ... \qquad [2.10]$$

with

$$\alpha = \frac{1}{4}(1 - \frac{\cos 2\Psi}{2}) \quad \beta = \frac{1}{4}(1 - \cos 2\Psi) \text{ and } \chi = \frac{1}{4}(\sin\Delta\sin 2\Psi) \qquad [2.11]$$

where β and χ are the Fourier coefficients of the sinusoidal function at double frequency 2f. Thus, the ellipsometric parameters can be determined from these coefficients.

2.5. Analysis of experimental data

Ellipsometry does not give results directly but if the physics of the interaction of light and matter are correctly introduced in a model, then the parameters characterizing a sample can be determined by fitting the theoretical results to the experimental values [DAH 04a,

DAH 04b, NOU 07, LOU 08]. For data analysis, an inverse method is applied from models built for the various layers. It is necessary to consider both the material structure and an optical model (e.g. Drude Lorentz or Cauchy, etc.) for the optical properties of the material (Figure 2.4). The structural model allows the study of the surface and interfaces of the samples and in particular polymer-substrates interfaces, corresponding to the different types of substrates present in a mechatronic module. The optical constants which describe the interaction of light (dispersion, absorption, scattering, reflection, etc.) of these layers allow the characterization of polymer substrate interfaces.

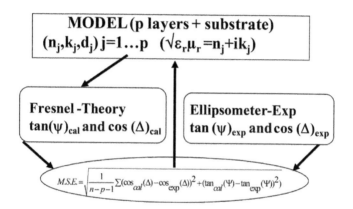

Figure 2.4. *Nonlinear regression method*

To determine the parameters of a thin layer or film (optical index n_f, extinction coefficient k_f and thickness e) on a substrate (optical index n_s and extinction coefficient k_s) from experimental parameters, with ψ and Δ supplied by the ellipsometer, a numerical procedure is built from a given algorithm, either simplex, Levenberg–Marquardt or Broyden–Fletcher–Goldfarb–Shanno [NEL 65, PRE 86, MAR 63, MOR 77]. For the system studied, ψ and Δ are the functions of all the parameters in the form:

$$\Psi = f(n_a, n_s, k_s, n_f, k_f, e_f) \qquad [2.12]$$

and

$$\Delta = g(n_a, n_s, k_s, n_f, k_f, e_f) \quad [2.13]$$

where the subscripts a, s and f correspond to the ambient medium, the substrate and the film.

The analysis consists of comparing the values of parameters $\tan(\psi)$ or ψ and $\cos(\Delta)$ or Δ calculated from a model with measured $\tan(\psi)$ and $\cos(\Delta)$ values as shown in Figure 2.4.

The adjustment is done by modifying the various optical parameters and thickness values on which the ellipsometric parameters depend. The calculated and experimental curves which are dependent on the light wavelength are fitted by minimizing an objective function or cost function mean square error (MSE) that estimates the difference between calculated and measured values by:

$$M.S.E. = \sqrt{\frac{1}{n-p-1} \sum_{1}^{n} (\cos_{cal}(\Delta) - \cos_{exp}(\Delta))^2 + (\tan_{cal}(\Psi) - \tan_{exp}(\Psi))^2)} \quad [2.14]$$

or, to take into account the fact that the signal-to-noise ratio is different in the spectral range of measurement and to better fit the experimental data polluted by noise, by:

$$M.S.E. = \sqrt{\frac{1}{n-p-1} \sum_{i=1}^{i=n} ((\frac{\Psi_i^{cal} - \Psi_i^{exp}}{\sigma_{\Psi,i}^{exp}})^2 + (\frac{\Delta_i^{cal} - \Delta_i^{exp}}{\sigma_{\Delta,i}^{exp}})^2))} \quad [2.15]$$

where n is the number of data points which are twice the number of wavelengths and/or incident angles selected for the measurements, p is the number of unknown parameters to be adjusted and σ is the standard deviation of the measured data. This equation has n parameters for $n/2$ wavelengths because there are two measured values included in the calculation for each pair of parameters ψ and Δ. Weighting the data by the inverse of the standard deviations reduces the contribution of the measurements polluted by noise to the adjustment. In the case of a film on a substrate, the medium is typically ambient air ($n_{air} = 1$) of a known index so that there are only five unknowns to be determined ($p = 5$) by the fitting algorithm (Simplex, Levenberg–Marquardt, etc.).

Thus, if the MSE decreases at each iterative step, it can be assumed that there is a tendency toward a fairly decent model fit. It should be noted that there is a strong correlation between the refractive index and the thickness. To reduce this interdependence, it is necessary to have a sufficient number of measurements depending on the number of parameters to be adjusted. Data from multiple wavelengths (spectroscopic ellipsometer) and/or angles of incidence (variable angle ellipsometer) can thus be combined if there are many parameters to adjust. For example, to fit three parameters, it takes at least six sets of measurements which can be in terms of three angles of incidence or three wavelengths for one angle of incidence.

2.6. The stack structural model

To illustrate analysis by the inverse method, an example of a thin film on a given substrate as shown in Figure 2.5 is considered. As surfaces are not perfectly plane, the surface roughness must also be taken into account in the model. The film is built as a stack of two flat and homogeneous layers of different materials and indices. The first layer essentially consists of the material whose index is to be determined (film) and depends on the parameters e_f, n_f and k_f. Above, the second layer is composed of an effective medium, corresponding to a mixture of "air-material", characterized by an effective index and a percentage of air included in the material to model the presence of roughness (effective medium).

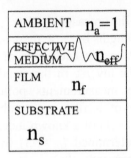

Figure 2.5. *Model of a thin layer on a substrate with a rough surface*

The second layer is characterized by a roughness of average thickness (e_r), as well as by a fraction of air inclusion in the layer (Maxwell–Garnett, Bruggemann or Philips model) and an index calculated by:

$$\frac{n_e^2 - n_h^2}{n_e^2 + 2n_h^2} = \sum_i f_i \frac{n_i^2 - n_h^2}{n_i^2 + 2n_h^2} \qquad [2.16]$$

where n_i is the complex index of the inclusion, n_h is the complex index of the host medium, n_e is the complex index of the effective medium and f_i is the fraction of inclusion. In general, the indices of the substrate and air are known, while all the other parameters (e_f, n_f, k_f, n_i, $k_i = 0$, $n_h = n_f$, $k_h = k_f$, n_e, k_e and e_r) characterizing the film and the rough layer are to be fitted to the measurements. The model can be refined by taking into account the inclusion of an intermediate layer between the substrate and the film.

2.7. The optical model

The optical model that characterizes a polycrystalline solid material is a function of its properties: metallic, insulating, semiconductor or semi-metal. A metal is characterized by a partially filled conduction band, while in the case of an insulator, this band is empty. In a semiconductor, the ambient temperature is sufficient to populate the conduction band. The nature of electrons (free, bound or some free and others bound) has led to different optical models, the most commonly used being the Cauchy or Sellmeir models for low-absorbing insulators, the Drude model for metals or the Lorentz model for intermediate cases. The Lorentz model is a parametric model because the transitions between two energy bands are modeled by a set of discrete transitions that do not correspond to the actual band structure.

The complex dielectric constant of a metal characterized by free electrons is calculated with the Drude model expressed as:

$$\varepsilon = 1 - i\frac{\sigma}{\omega \varepsilon_0} = 1 - \frac{N_e e^2 / m\varepsilon_0}{\omega^2 - i\omega \Gamma_D} \quad [2.17]$$

where ε_0 is the dielectric constant of vacuum, σ is the conductivity of the metal, N_e is the density of the electrons contributing to σ and ω is the angular frequency of the wave; calling ε_∞ the dielectric constant at high frequency, this equation becomes:

$$\varepsilon = \varepsilon_\infty - \frac{\omega_p^2}{\omega^2 - i\omega \Gamma_D} \quad [2.18]$$

where ω_p is the plasma frequency which is given by:

$$\omega_p = \sqrt{N_e e^2 / m\varepsilon_0} \quad [2.19]$$

and where $\omega\Gamma_D$, the imaginary part of the dielectric constant, corresponds to the absorption which is proportional to the time average of optical electron collisions. The term ε_∞ represents the contribution of the lattice and the second term represents the contribution of the free electrons.

From this model, the electrical resistivity ρ_e can be calculated by:

$$\rho_e = \frac{1}{2\pi\varepsilon_0 c} \frac{\Gamma_D}{\omega_p^2} \quad [2.20]$$

where c is the velocity of light.

The second mechanism, called inter-band, is related to optical transitions between two electronic bands, from full (valence) bands to the states of the conduction band or from the latter to empty higher energy bands. Its contribution to the dielectric constant is described by the Lorentz model. In this way, we obtain for the complex dielectric constant:

$$\varepsilon = 1 + \sum_i \frac{N_e e^2/m\varepsilon_0}{\omega_i^2 - \omega^2 - i\omega\Gamma_0} \qquad [2.21]$$

where N_e is the electronic density contributing to the transition. Calling ε_∞ the dielectric constant at high frequencies, we can cast this equation as:

$$\varepsilon = \varepsilon_\infty + \sum_i \frac{N_e e^2/m\varepsilon_0}{\omega_i^2 - \omega^2 - i\omega\Gamma_0} = \varepsilon_\infty + \sum_{i=1}^{2} \frac{\omega_p^e}{\omega_i^2 - \omega^2 - i\omega\Gamma_0} \qquad [2.22]$$

where

$$\omega_p^e = \sqrt{N_e e^2/m\varepsilon_0} \qquad [2.23]$$

ω_i is the central frequency of the oscillator and Γ_0 is the broadening of the oscillator. The term ε_∞ represents the contribution of the lattice and the second term represents the contribution of two oscillators. The Drude–Lorentz model allows us to retrieve the optical properties from:

$$\varepsilon = \tilde{n}^2 \qquad [2.24]$$

where:

$$\varepsilon = \varepsilon_r + i\varepsilon_i \quad \text{and} \quad \tilde{n}^2 = (n - ik)^2 \qquad [2.25]$$

The following relationships between the real and imaginary parts of the dielectric function and the complex optical constant can thus be deduced as:

$$\varepsilon_r = n^2 - k^2 \quad \text{and} \quad \varepsilon_i = 2nk \qquad [2.26]$$

The optical constants can be determined by inverting the formulas, which leads to:

$$n = \sqrt{\frac{\varepsilon_r + \sqrt{\varepsilon_r^2 + \varepsilon_i^2}}{2}} \text{ and } k = \sqrt{\frac{-\varepsilon_r + \sqrt{\varepsilon_r^2 + \varepsilon_i^2}}{2}} \quad [2.27]$$

2.8. Application of ellipsometry technique

In this section, the results obtained on a thin layer of nanograins of silver sintered on a copper substrate and a polymer-based material on a metallic or silicon substrate are given. The results are shown and discussed for models consisting of thin films on various substrates as a function of temperature in a wet and a dry environment in terms of optical properties.

The measurements were performed with the variable angle spectroscopic ellipsometer, M-2000V, from the J.A. Woollam Company [WOO 00, DAH 10, ALA 11, SCI 12, KHE 14]. The M-2000V is a rotating compensator (RCE) with a charged-coupled device (CCD) detection system. The light source is a tungsten lamp emitting in a spectral range from 370 to 1,000 nm. The apparatus is manually operated for different steps such as the calibration check, beam alignment and data acquisition. To maximize the signal-to-noise ratio, a 100 s measuring time duration is chosen for each spectrum.

Studies were performed for an incident angle of 70° (at this value, the dimensions of the spot of light reflected by the sample are about 10 mm × 3 mm, representing a surface which is smaller than that of the substrate). It is then possible to perform measurements at the center of the sample, over an area where the thickness is thin and fairly uniform. A thermal cell associated with the ellipsometer makes measurements according to the temperature possible. An example of a recorded spectrum is given in Figure 2.6. SE measurements are then analyzed using the Woollam Complete EASE software from models available in the database. The physical properties of each layer in a stack of layers deposited on a substrate can also be described by corresponding parametric optical models from a database as shown in Figure 2.7. On top of these layers, a final one can be added to simulate the roughness of the sample's surface. If the model is representative of the sample, the theoretical ellipsometric parameters will be in good agreement with the measured data, as shown by the correspondence between the dotted lines and the curves in Figure 2.7.

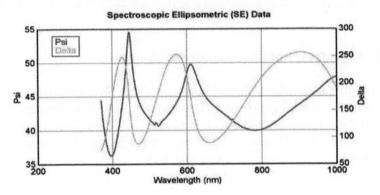

Figure 2.6. *Ellipsometric parameters. For a color version of this figure, see www.iste.co.uk/elhami/embedded1b.zip*

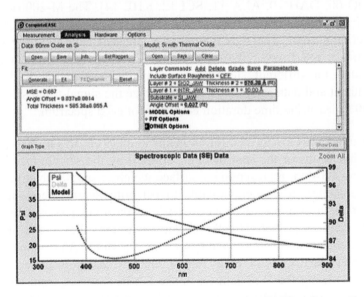

Figure 2.7. *Example of model and fit using CompleteEASE software. For a color version of this figure, see www.iste.co.uk/elhami/embedded1b.zip*

2.8.1. *Thin layer from silver nanograins sintered on a copper substrate*

The silver sample was prepared by spark plasma sintering (SPS) under pressure (33 MPa) [DAH 10, ALA 11] on copper. Ellipsometric studies were performed as a function of the temperature between –30 and 250°C. The structural model is composed of a copper (Cu) substrate coated with a layer of silver without roughness. Two types of silver layer were chosen for the analysis: one modeled by the Drude dielectric function (Ag-Gen-Osc function) and another modeled as an effective medium coupled ("EMA-coupled") layer.

In this case, the dielectric Lorentz function is added (Ag-Lorentz) to the Drude dielectric function. In the framework of the Drude model, the electrons are delocalized (or free) and the parameters to be adjusted are the resistivity (rho: ρ ($\mu\Omega$.cm)) and the time constant relaxation (tau: τ (fs)). Reference values were taken from [VIA 07] as starting values for the fit. With the Lorentz model, which models the localized (or bound) electrons, the parameters are the amplitude of an oscillator (the intensity of the line i: A_i (eV)), the coefficient of broadening (the linewidth i: B_i (eV)) and the frequency of the oscillation (the center line i: E_i (eV)). Two oscillators were used and the initial values are taken from [VIA 07]. The analysis was performed in two steps, first on the entire spectral range from 370 to 1,000 nm for the following temperatures: [–30, 0, 25, 30, 60, 62, 72, 82, 92, 113, 134, 150, 178, 206, 250]°C, and then within defined spectral regions at different temperatures. This analysis has, therefore, allowed us to determine the resistivity and the relaxation time constant as a function of the temperature, and the skin depth as a function of the wavelength for a given temperature. The results are shown in Figures 2.8–2.11.

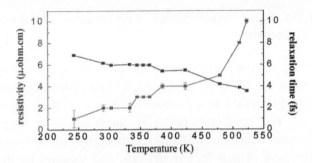

Figure 2.8. *Change of the resistivity ρ (μΩ.cm) and the mean time τ (fs) as a function of temperature. For a color version of this figure, see www.iste.co.uk/elhami/embedded1b.zip*

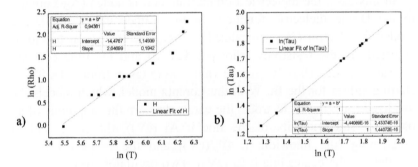

Figure 2.9. *Change as a function of temperature a) of the resistivity and b) of the relaxation time (log-log scale)*

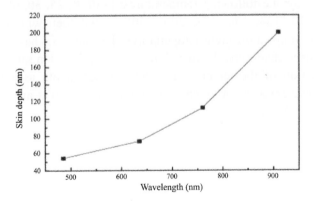

Figure 2.10. *Change in the skin depth as a function of the wavelength at a temperature of 25°C for fixed ρ*

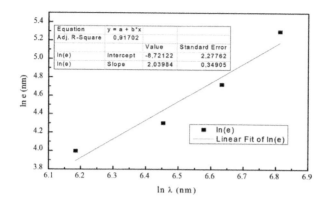

Figure 2.11. *Linear plot of the skin depth as a function of the wavelength (log-log scale)*

At low temperatures (−30°C), the spectra are noisy but we obtain a resistivity which varies around 3 µΩ.cm. For the other temperatures, the model used for fitting allows overlapping of the experimental and calculated spectra. There is an increase in resistivity when the temperature increases.

Plotting the logarithmic of ρ and τ against the logarithm of temperature T, we obtain the linear relationships shown in Figures 2.9(a) and (b), which allows extrapolation outside the experimental range.

The skin depth is related to the resistivity and the wavelength by the following theoretical formula:

$$e = \sqrt{\frac{\rho \lambda}{\pi c \mu_0 \mu_r}} \quad [2.28]$$

By applying the logarithmic function to relationship [2.28], a linear function is obtained as: ln e = const. + 1/2 ln (λ), which allows us to draw the straight line shown in Figure 2.11, thus validating the values determined experimentally by SE.

2.8.2. Analysis of ellipsometric spectra of polymers on different substrates

The study is presented for seven samples of silicone gels and epoxy resin, deposited by the technique of spin coating on different substrates: quartz, aluminum, nickel-plated copper and silicon. The samples are denoted as B1, B2, B3 and B4 for the bi-component, M1, M2 and M3 for the mono-component and finally E1 for the epoxy resin.

The quartz substrate is selected for its temperature stability and chemical inertness with respect to the polymers in order to allow the creation of a database of optical constants to be used as a reference to interpret results of the measurements with the other substrates that have materials present in an insulated molded frame (IML)-type model. The samples on the quartz substrates were used to determine the values of the refractive index n following two operating procedures to analyze the results, called protocol 1 (material and substrate) and protocol 2 (material considered as substrate). For the other samples, a first ellipsometric study was conducted at room temperature (T = 25°C) at an incident angle i of 70°, and then a second study was conducted after they were subjected to thermal stress in a humid atmosphere (T+H).

Figure 2.12 shows the ellipsometric parameters Psi (upper curve) and Delta (lower curve) measured experimentally on the material B1, and the curves fitted (dashed lines) with the B-spline (mathematical method) model from the software Complete EASE [SCI 12, KHE 14]. The MSE of 1.5 indicates good agreement between the experimental values and calculations as shown in the overlapping of the measured and fitted curves in Figure 2.12.

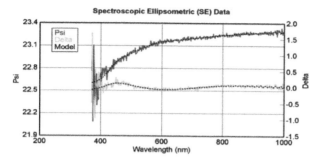

Figure 2.12. *Ellipsometric parameters, Psi and Delta, measured and fitted. For a color version of this figure, see www.iste.co.uk/elhami/embedded1b.zip*

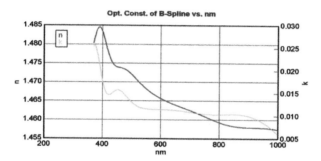

Figure 2.13. *Refractive index n and extinction coefficient k of material B1 after fit with B-spline model. For a color version of this figure, see www.iste.co.uk/elhami/embedded1b.zip*

Figure 2.13 shows the optical constants, n and k. The refractive index n decreases with the wavelength from 1.48 to 1.46 between 425 and 950 nm, while the absorption k varies from 0.026 to 0.01. This decrease is not monotonous throughout the interval, and below 425 nm, the signal-to-noise ratio does not allow a correct fit. It is further noted that the material is slightly absorbing between 450 and 700 nm.

With quartz as the substrate, a change of optical constants like that with B1 is determined with all the samples. The values of n are between 1.45 and 1.48 for B1, B2, B3, B4 and M1, between 1.36 and 1.38 for M2 and between 1.38 and 1.44 for M3, and it remains around 1.5 for the epoxy resin. Regarding k, the values are still very low, below 10^{-2} (B1 10^{-2}; B2 between 10^{-4} and 2.5×10^{-2}; B3 between 10^{-4} and 3×10^{-3}; B4 between 10^{-4} and 8×10^{-3}, M1 from 5×10^{-3} to 3×10^{-2}; M2 from 10^{-3} to 6×10^{-2}; M3 from 10^{-2} to 7×10^{-2}; E1 from 10^{-2} to 5×10^{-2}); the comparison of the values determined with the physical model of Cauchy and the mathematical B-spline method shows that we obtain values that are very close for n and k. By way of example, we give the results obtained with B1, which are shown in Figures 2.14(a) and (b). Both models give comparable results, the mean square deviation between the two sets of measurements being 10^{-6} and 7×10^{-7} for n and k, respectively.

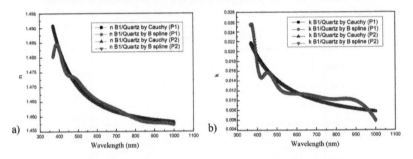

Figure 2.14. *Comparison of the optical constants of the polymer B1 obtained with the B-spline and Cauchy models: a) refractive index and b) extinction coefficient. For a color version of this figure, see www.iste.co.uk/elhami/embedded1b.zip*

With the P2 (substrate model) protocol, we obtain values of n and k that are identical to those given by P1, but the adjustment does not reproduce the oscillations, the latter being mainly related to the thickness of the material. Analyses show that for thicknesses that are greater than 5 microns, both protocols lead to the same optical constant values.

If we are only interested in these constants, we can perform the analysis with the P2 protocol. Indeed, for an incident angle i of incidence 70°, a thickness of 5 µm is equivalent to a path of 12 µm inside the material, and so this value is about 20 times the average wavelength, which is around 600 nm. In this case, light is not very sensitive to inhomogeneities in the material–substrate interface.

If the thickness is of the same order of magnitude as the wavelength, the protocol P2 can be used to determine the optical constants.

Analyses were performed by first determining the values of the thickness, the dielectric function and the other adjustment parameters (roughness and offset angle) as a function of the wavelength with the "B-spline model". The thickness is fixed until the simulations reproduce the oscillations present in the spectra. The B-spline model is then replaced by the Cauchy model.

The comparison of the refractive index and the extinction coefficient of the material B1 deposited on different substrates is shown in Figures 2.15(a) and (b). Given the large values of n for B1 on Si, the corresponding variation is given in the insert.

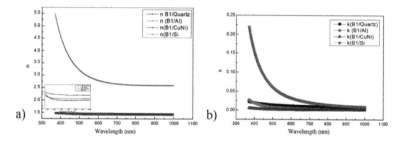

Figure 2.15. *Comparison of a) n and b) k, of polymer B1 on different substrates before heat treatment in a humid atmosphere. For a color version of this figure, see www.iste.co.uk/elhami/embedded1b.zip*

The refractive index n of the material B1 on different substrates decreases as the wavelength increases. In the case of silicon, we note a variation from 5.5 to 2.5 between 400 and 750 nm and a plateau at about 2.5 between 750 and 1,000 nm. In the case of the other substrates, n remains near 1.5. The sharp variation of n for the silicon corresponds to an area of strong absorption resulting in a decrease of k from 0.2 to 10^{-3} between 370 and 900 nm. For the other substrates, k varies very slightly between 400 and 650 nm and goes to 0 beyond 650 nm. The comparison shows that the index of refraction of the material varies on the same order of magnitude from 1.38 to 1.48 for quartz, CuNi and Al substrates, with roughly the same k values. However, on silicon, the refractive index takes other values between

2.5 and 5.5. This sharp difference is synonymous with a significant change that occurred upon deposition. The hypothesis of a possible interaction between silicone polymer B1 and silicon seems to be a plausible cause in this case.

In this respect, the change of the refractive index of polymer B1 deposited on silicon can reasonably be attributed to a chemical interaction which completely modifies the initial molecular lattice to give a new material with different properties. For this reason, the index of the polymer on silicon takes values that are very different from 1.5. Similar results are found for all polymers with a smaller effect for M3 (Figure 2.16). In the case of E1 (Figure 2.17), we can infer that there was no interaction since n varies little from 1.53 to 1.56, remaining close to 1.5. As shown in the curves in Figure 2.14 for B1 and Figure 2.15 for M3, similar results were obtained for the other polymers of type B (B2, B3, B4) and of type M (M1, M2), respectively.

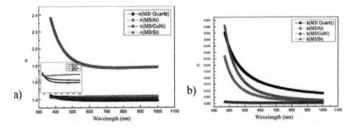

Figure 2.16. *a) Refractive index and b) extinction coefficient of polymer M3, on different substrates (Si, CuNi, Al and quartz). For a color version of this figure, see www.iste.co.uk/elhami/embedded1b.zip*

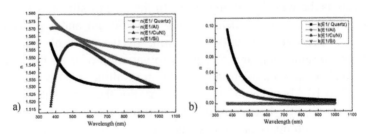

Figure 2.17. *a) Refractive index and b) extinction coefficient of epoxy resin E1 on different substrates (Si, CuNi, Al and quartz). For a color version of this figure, see www.iste.co.uk/elhami/embedded1b.zip*

2.8.3. Analysis and comparison after stress

The same methodology was used to characterize the materials deposited on the various substrates that were heated in a humid atmosphere. The results obtained before and after stress are shown in Figure 2.18 for B1.

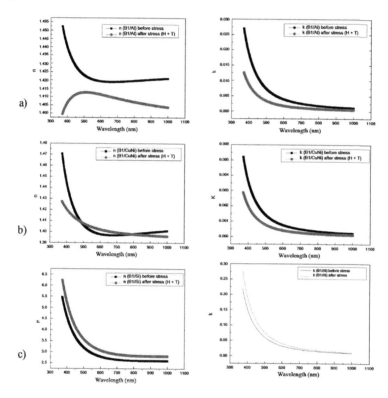

Figure 2.18. *Comparison of the refractive index n and the extinction coefficient k of materials B1/Al, B1/CuNi and B1/Si before and after stress. For a color version of this figure, see www.iste.co.uk/elhami/embedded1b.zip*

The consequence of the environmental constraint imposed on the materials is the modification of their refractive index n which is observed for almost all the materials regardless of the substrate, except for the M1/CuNi, M3/Si and B2/Si samples. Note that on nickel-copper, variations are small compared to those recorded on aluminum, for which even the shapes of the curves were altered.

During deposition, the bonds formed between the silicon and the silicone gel irreversibly modified the latter's optical properties since the refractive index does not undergo any significant variation compared to that determined in the case of the other substrates (CuNi and Al). Note that the differences are quite significant on the aluminum and nickel-plated copper substrates. Variations are recorded for k for all materials, except for the sample M3/Si. On the Si substrate, these variations are small. For the two materials M1 and B2 on the CuNi substrate, $k = 0$, thus showing that these materials are transparent. No changes are observed in the curves obtained for the silicon substrates, suggesting that the optical properties of the new material obtained in contact with Si are stable. Also, note that on CuNi, just as for n, variations in k are less significant compared to those recorded with Al substrate, for which even the shape of the curves changed.

2.8.4. Physical analysis of light and matter interaction in terms of band gap energy

Ellipsometric spectra in the UV-visible range allow the study, from the energy of the band gap, of the effect of an environmental stress on the structure of the material [MIR 13]. By studying the variation of the energy of the band gap (E_g), as a function of the substrates, we can indirectly determine the effect of the substrate on the polymer at the interface (polymer/substrate). Below, we develop the results of this study in the case of material B1 on quartz and aluminum.

2.8.4.1. Analysis of the B1/quartz interface

The value of the energy of the bandgap (E_g) can be obtained from the absorption curve (Figure 2.19).

From the following formula:

$$\alpha h\upsilon \sim (h\upsilon - E_g)\, n/2 \text{ with } n = 1 \text{ or } 2 \qquad [2.29]$$

where, depending on whether the transition between the valence band and the conduction band is direct or indirect (n = 1 or 2, respectively), we can determine E_g. As shown in Figure 2.19, the value of the band gap energy (E_g) is equal to 3 eV after extrapolation in the case of B1 on quartz.

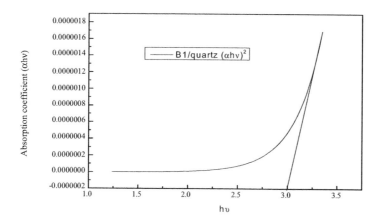

Figure 2.19. *Analysis of material B1/quartz and determination of the gap energy*

2.8.4.2. *Analysis of the B1/aluminum interface*

Figures 2.20(a) and (b) show the fit of the experimental data and the parameters of the physical model "GenOsc" from the Complete EASE software in the UV-visible region. Two oscillators of the Cody-Lorenz type were used. In that case, the band gap energy is given by the imaginary part of the dielectric function as:

$$\varepsilon_2(E) = \alpha(E - E_g)^2 \qquad [2.30]$$

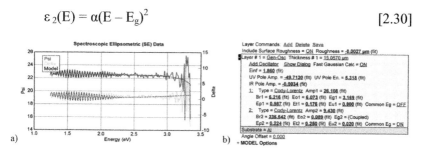

Figure 2.20. *Analysis of interface B1/Al: a) ellipsometric parameters and b) GenOsc Model for the fit. For a color version of this figure, see www.iste.co.uk/elhami/embedded1b.zip*

An energy gap $E_g = 3.169$ eV was determined. The gap increases from the quartz substrate to the aluminum substrate. This result can be interpreted on the basis of a change in the lattice parameters of the system in contact with the substrate, which are higher in the case of quartz than in the case of aluminum.

2.9. Conclusion

Spectroscopic ellipsometry (SE) is a non-destructive characterization technique for surfaces and interfaces which can be used to study defects caused by environmental stresses, such as the effect of temperature on the transport properties (electrical and thermal conductivity) of sintered Ag samples.

Two protocols (P1 and P2) can be used to study thick transparent materials. P1 is used for thin layers (silicon gel and epoxy resin) deposited on various substrates (coating thickness less than 5 µm), and P2 is used for thick substrates. Studies of the effect of thermal stresses on the interfaces of polymers deposited on silicon substrates reveal changes in the material characteristics from the refractive index. Only one polymer (M3) shows no chemical reactivity. Substrates that have reacted with the polymer coating, in this case silicon, are identified.

SE was applied to study the effect of thermal stresses in a moist environment on the optical properties of various coated samples. Important changes in optical properties were detected on aluminum substrates but not on copper-nickel substrates. No change was observed when the polymer reacted with the substrate (silicon) when the sample was prepared before being submitted to environmental stresses.

The results of this chapter show that the interfaces between the various materials present in a power module can be altered by environmental stresses (temperature, humidity, vibration, etc.). This modification, which depends on the physical and chemical properties of the materials in contact, can be the cause of failures in mechatronic modules. For example, if the connection wires are made of aluminum, it is likely that in a humid atmosphere, protection with a polymer coating will not be as effective as with copper wires because the interface will be modified as observed by SE. Study by SE thus allows the anticipation of the probable causes of failures on the basis of a good modeling of the interfaces.

2.10. References

[ALA 11] ALAYLI N., Frittage de pâte de nano et micro grains d'argent pour l'interconnexion dans un module de mécatronique de puissance : Elaboration, caractérisation et mise en œuvre, PhD thesis, UVSQ, Versailles, 2011.

[AZZ 77] AZZAM R.M.A., BASHARA N.M., *Ellipsometry and Polarized Light*, North Holland Co., Amsterdam, 1977.

[BOC 93] BOCCARA A.C., PICKERING C., RIVORY J. (eds), *Spectroscopic Ellipsometry Proceedings*, Elsevier, Amsterdam, 1993.

[DAH 03] DAHOO P.R., HAMON T., SCHNEIDER M. *et al.*, "Ellipsometry: Principles, signal processing and applications to metrology", *Proceedings de la Conférence Internationale sur la Modélisation Numérique Appliquée CIMNA*, Beirut, Lebanon, 14–15 November, 2003.

[DAH 04a] DAHOO P.R., HAMON T., NEGULESCU B. *et al.*, "Evidence by spectroscopic ellipsometry of optical property change in pulsed laser deposited NiO films when heated in air at Neel temperature", *Applied Physics A: Materials Science and Processing*, vol. 79, pp. 1439–1443, 2004.

[DAH 04b] DAHOO P.R., GIRARD A., TESSEIR M. *et al.*, "Characterizaton of pulsed laser deposited SmFeO3 morphology: Effect of fluence, substrate temperature and oxygen pressure", *Applied Physics A: Materials Science and Processing*, vol. 79, pp. 1399–1403, 2004.

[DAH 10] DAHOO P.R., ALAYLI N., GIRARD A. *et al.*, "Reliability in mechatronic systems from TEM, SEM and SE material analysis", *Material Research Society Proceedings*, vol. 1195, pp. 183–190, 2010.

[DRE 82] DREVILLON B., PERRIN J., MAROT R. *et al.*, "Fast polarization modulated ellipsometer using a microprocessor system for digital Fourier analysis", *Review of Scientific Instruments*, vol. 53, p. 969, 1982.

[DRU 87] DRUDE P., "By the laws of reflection and refraction of light at the boundary of absorbing crystals", *Annals of Physics*, vol. 32, p. 584, 1887.

[FUJ 07] FUJIWARA H., *Spectroscopic Ellipsometry, Principles and Applications*, John Wiley & Sons, 2007.

[KHE 14] KHETTAB M., Etude de l'influence du résinage au niveau de L'IML (Insulated Metal Leadframe), dans le packaging de module commutateur de courant mécatronique, PhD thesis, UVSQ, Versailles, 2014.

[MAR 63] MARQUARDT D.W., "An algorithm for least squares estimation of nonlinear parameters", *Society for Industrial and Applied Mathematics*, vol. 11, p. 431, 1963.

[MIR 13] MIR F.A., BANDAY J.A., CHONG C. et al., "Optical and electrical characterization of Ni-doped orthoferrites thin films prepared by sol-gel process", *The European Physical Journal Applied Physics*, vol. 61, pp. 10302–10305, 2013.

[MOR 77] MORE J., "The Levenberg-Marquardt algorithm: Implementation and theory", in WATSON G.A. (ed.), *Numerical Analysis*, Springer-Verlag, 1977.

[NEL 65] NELDER J.A., MEAD R., "A simplex method for function optimization", *Computer Journal*, vol. 7, p. 308, 1965.

[NOU 07] NOUN W., BERINI B., DUMONT Y. et al., "Correlation between electrical and ellipsometric properties on high-quality epitaxial thin films of the conductive oxide LaNiO3 on STO (001)", *Journal of Applied Physics*, vol. 102, pp. 063709-1–063709-7, 2007.

[OHL 00] OHLIDAL I., FRANTA D., *Progress in Optics*, Elsevier, 2000.

[PIC 95] PICKERING C., *Photonic Probes of Surfaces*, Elsevier, 1995.

[PRE 86] PRESS W.H., TEUKOLSKY S.A., VETTERLING W.T. et al., *Numerical Recipes*, Cambridge University Press, 1986.

[ROT 45] ROTHEN A., "The ellipsometer, an apparatus to measure thicknesses of thin surface films", *The Review of Scientific Instruments*, vol. 16, p. 26, 1945.

[SCI 12] SCIAMMA-O'BRIEN E., DAHOO P.R., HADAMCIK E. et al., "Optical constant from 370 nm to 900 nm of Titan tholins produced in a low pressure RF plasma discharge", *Icarus*, vol. 218, pp. 356–363, 2012.

[TOM 99] TOMPKINS H.G., MCGAHAN W.A., *Spectroscopic Ellipsometry and Reflectometry*, John Wiley & Sons, New York, 1999.

[TOM 05] TOMPKINS H.G., IRENE E.A., *Handbook of Ellipsometry*, William Andrew Inc., Springer, New York, 2005.

[VED 98] VEDAM K., "Spectroscopic Ellipsometry: A historical overview", *Thin Solid Films*, vol. 313, p. 1, 1998.

[VIA 07] VIAL A., LAROCHE T., "Comparison of gold and silver dispersion laws for FDTD simulations", *Journal of Physics D: Applied Physics*, vol. 40, p. 7152, 2007.

[WOO 00] WOOLLAM J.A., "Ellipsometry, variable angle spectroscopic", in WEBSTER J.G. (ed.), *Encyclopedia of Electrical and Electronics Engineering*, John Wiley & Sons, New York, 2000.

3

Method of Characterizing the Electromagnetic Environment in Hyperfrequency Circuits Encapsulated Within Metallic Cavities

The designers of hyperfrequency modules accord particular importance to electromagnetic compatibility. To protect hyperfrequency power modules from external electromagnetic disturbances, such modules are enclosed inside metallic cavities. This prevents them from disturbing the neighboring circuits. In radar applications, power module output can exceed hundreds of kilowatts so their radiated emissions are very high. The harsh electromagnetism (EM) environment found in enclosed power modules can lead to failure mechanisms in the power transistors and impact on the reliability of radar systems. It is thus important to understand and control the EM environment of enclosed hyperfrequency power modules. In this chapter an innovative concept of determining the emissions radiated by the hyperfrequency structures found in metallic cavities is presented. This method is based on near-field measurements obtained from non-enclosed units and on radiated emission modeling by a network of dipoles.

3.1. Introduction

In this chapter, we present reliability research in gallium nitride (GaN) semiconductor technologies used in radar modules [KHE 11]. These modules are enclosed in metallic cavities. Several active

---

Chapter written by Samh KHEMIRI, Abhishek RAMANUJAN, Moncef KADI and Zouheir RIAH.

components processing high-frequency signals are located in the same housing. This encapsulation leads to a very harsh electromagnetic environment which can cause degradation in the structure of the active and passive components.

In a reliability study, it is important to analyze the actual electromagnetic environment of the component of the GaN technology. This allows us to understand the real operating conditions of the transistor under test. In this chapter, first of all we will study the effect of the metallic cavity on the distribution of the electromagnetic field radiated by a hyperfrequency circuit (e.g. a high-power amplifier (HPA)) for various operating frequencies. Second, we will develop a method for determining the electromagnetic environments of radio frequency (RF) power modules enclosed within metallic cavities. This consists of using the knowledge of the electromagnetic environment of uncovered modules, using near-field measurements, in order to estimate the electromagnetic environment with enclosed modules.

3.2. Theory of metallic cavities

3.2.1. *Definition*

An electromagnetic cavity is a closed volume filled with one or more dielectric materials and is limited by electric or magnetic walls. In the case where the cavity is surrounded by conductors, it is called a metal cavity and in the other case, the cavity is a dielectric resonator [COM 96, KON 75, ARC 03]. Generally, the cavities may have any geometric shape but, in practice, they have a cylindrical shape with a circular or rectangular section.

The metal cavities can be obtained by closing rectangular or circular wave guides with the metallic plates, perpendicular to their longitudinal axis.

3.2.2. Electromagnetic field in a parallelepiped cavity

The theoretical distribution of the electromagnetic fields in a parallelepiped cavity of dimensions a, b and c with $a < b < c$ is given by the following equations:

$$E_x = A_1 \cos\left(\frac{m\pi}{a}.x\right).\sin\left(\frac{n\pi}{b}.y\right).\sin\left(\frac{p\pi}{c}.z\right) \qquad [3.1]$$

$$E_y = A_2 \cos\left(\frac{n\pi}{b}.y\right).\sin\left(\frac{p\pi}{c}.z\right).\sin\left(\frac{m\pi}{a}.x\right) \qquad [3.2]$$

$$E_z = A_3 \cos\left(\frac{p\pi}{c}.z\right).\sin\left(\frac{n\pi}{b}.y\right).\sin\left(\frac{m\pi}{a}.x\right) \qquad [3.3]$$

$$H_x = B_1 \sin\left(\frac{m\pi}{a}.x\right).\cos\left(\frac{n\pi}{b}.y\right).\cos\left(\frac{p\pi}{c}.z\right) \qquad [3.4]$$

$$H_y = B_2 \sin\left(\frac{n\pi}{b}.y\right).\cos\left(\frac{p\pi}{c}.z\right).\cos\left(\frac{m\pi}{a}.x\right) \qquad [3.5]$$

$$H_z = B_3 \sin\left(\frac{p\pi}{c}.z\right).\cos\left(\frac{m\pi}{a}.x\right).\cos\left(\frac{n\pi}{b}.y\right) \qquad [3.6]$$

where:

– A_1, A_2, A_3, B_1, B_2 and B_3 are constants;
– m, n and p are the indices of the normal modes of the cavity.

3.2.3. Resonance frequencies

A metallic cavity of dimensions (a, b, c) (Figure 3.1) stores energy for a discrete set of frequencies, called resonance frequencies, to which correspond particular configurations of the electromagnetic fields, called modes. These frequencies are given by the following expression:

$$f_{m,n,p} = \frac{C_0}{2\sqrt{\mu_r \varepsilon_r}} \sqrt{\left(\frac{m}{a}\right)^2 + \left(\frac{n}{b}\right)^2 + \left(\frac{p}{c}\right)^2}$$ [3.7]

where:

– C_0 is the speed of light (3×10^8 m/s) and (m, n, p) are the indices of the modes;

– μ_r is the permeability and ε_r is the permittivity of the substrate.

Expression [3.7] is only valid when the cavity is composed of a single dielectric. Otherwise, in the case where the hyperfrequency circuits are built on substrates of permittivity between 3 and 10, the cavity is partially filled by air and the substrate, we need to replace the permittivity, ε_r, by an effective equivalent permittivity.

In the literature, there are several formulas which can express this equivalent permittivity, but the most commonly used one is given by:

$$\varepsilon_{r_{équi}} = \frac{\varepsilon_r \cdot (h_{substrat} + h_{air})}{\varepsilon_r \cdot h_{substrat} + h_{air}}$$ [3.8]

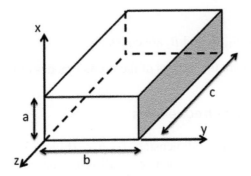

Figure 3.1. *Metallic cavity*

The resonance modes of a cavity are called eigenmodes and are denoted by TM_{mnp} (the transverse magnetic modes denoted by TM_{mnp} or waves H such that $H_z = 0$) and TE_{mnp} (the transverse electric modes denoted by TE_{mnp}, or waves E such that $E_z = 0$). The indices m and n depend on the order of the considered mode TE or TM. These are two positive or zero integers, but they cannot both be simultaneously zero. The index p is always strictly positive. In this configuration, the mode TE_{011} is the dominant mode where the lowest frequency for which a resonance of the cavity can be stimulated [ARC 03].

3.3. Effect of metal cavities on the radiated emissions of microwave circuits

In this section, we will look at the study of the influence of the metallic cavity on the distribution of the electromagnetic field of a given circuit as a function of the operating frequency. For example, an HPA is composed of several active components emitting very high electromagnetic fields. These could make our analysis of the cavity effect very tricky. For reasons of simplification and in order to correctly identify the influence of the cavity, we have decided to use a simple case study: a microstrip line with the same transverse dimensions, substrate and back plate as the HPA module.

3.3.1. *Circuit case study: 50 Ω microstrip line*

The microstrip line is a device that is often used in microwave circuits. It is composed of a metal trace of width w, length l and thickness t that is built on a substrate with relative permittivity ε_r and height h. The substrate is backed with a metallic plate forming a ground plane.

For a microstrip line, the electromagnetic wave partly propagates in the substrate in a quasi-transverse electromagnetic (TEM) mode. However, some of the field lines are also found in the air region above the conductor. When the line is excited, an electric field and a magnetic field, which are perpendicular to each other, are created between the metallic parts and substrates. These two fields are perpendicular to the direction of wave propagation. The different field components, denoted as E_x, E_y, E_z, H_x, H_y and H_z according to the Cartesian coordinate system, can be measured using the near-field scan technique. This measurement is easily performed when the test structures are open but nearly impossible for enclosed structures. It is for this reason that we are looking, in this chapter, to deal with this problem by developing a method which uses the near-field measurements and cartographies of open structures in order to be informed about the new environment when they are encapsulated within metallic structures.

In microwave theory, the length of the printed circuit line (trace) can be equal to or higher than the wavelength. In these conditions, the value of the electric and magnetic fields of the wave can vary along the line. Depending on the characteristic impedance of the line and the input/output impedances, the electromagnetic waves can be subjected to multiple reflections. In such cases, standing waves are created. These are characterized by the appearance of nodes and antinodes of voltages and currents. Every voltage antinode is the source of maximum electric field radiation and every current antinode is the source of maximum magnetic radiation.

In our study, the microstrip line being considered presents a characteristic impedance of 50 Ω, a length l = 60 mm and a width w = 0.6 mm (Figure 3.2). It is designed on a substrate of Rogers DUROID 6010.2 type with a thickness of 0.635 mm, length 60 mm, width 30 mm and permittivity ε_r = 10.2. The substrate is mounted on an aluminum sole of thickness 4 mm acting as the ground plane. The line is protected by a cavity of width a = 10 mm, height b = 30 mm and length c = 60 mm. The distance between the shielding and the substrate is therefore equal to 5.3 mm.

3.3.1.1. *Effect on the S parameters*

Equation [3.8], which expresses the resonance frequency of the cavity, allows us to provide an infinity of resonance frequencies. From a practical point of view, the difficulty lies in knowing which of these frequencies is stimulated by the signal propagating on the line for a certain frequency range. For this purpose, we have drawn the parameters S_{21} on the frequency range (1–8 GHz) of the cavity line system following two different configurations: with and without shielding (in other words, we eliminate the upper surface of the cavity).

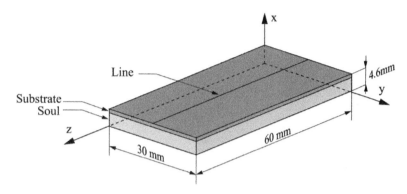

Figure 3.2. *Microstrip line*

According to the S_{21} curves shown in Figure 3.3, there is a significant variation between the two configurations. At two specific frequencies, 5.33 and 6.75 GHz, the signal propagating on the microstrip line is disturbed. At these frequencies, the S_{21} parameter decreases because a certain portion of the stored electromagnetic energy is reflected within the cavity [ARC 03]. These specific frequencies correspond to the eigenmodes of the cavity with indices (0, 1, 1) and (0, 1, 2) and are only excited by the microstrip line when it is encapsulated within the cavity in this frequency range. The excitation of resonant modes, in general, may also occur when the circuit has discontinuities such as open-circuit lines, short-circuit lines, gaps and sharp bends (90° bends) [JAN 89, DIX 05, FAN 94]. The stimulation of these modes also depends on the position of the line in the cavity; in other words, if the discontinuity is close to the metallic walls of the cavity, the stimulation is easier.

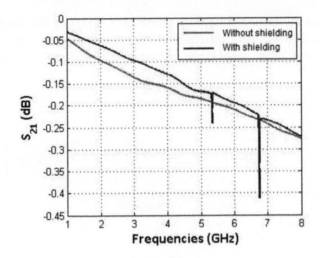

Figure 3.3. *Parameter S_1 of the system. For a color version of this figure, see www.iste.co.uk/elhami/embedded1b.zip*

3.3.1.2. *Effect on the magnetic field cartographies*

The electromagnetic field radiated by the line is calculated for the two configurations (open cavity and closed cavity) for a frequency of 3 GHz. The field line profiles are calculated using the three-dimensional (3D) electromagnetic simulator high frequency full wave electromagnetic field simulator (HFSS), based on the finite element method (FEM) in the frequency domain. The cartographies presented in this chapter are identified in the (*yoz*) plane located 2 mm above the microstrip line.

The distribution of the electromagnetic field for the two configurations, with and without shielding, for a frequency equal to 3 GHz (frequency in the range *S*), are displayed in Figures 3.4 and 3.5. No impact can be observed on the field profiles except in the neighborhood of the metallic linings. Indeed, the electric field E_z is subject to a decrease of 10 V/m. Moreover, the maximum values of all the components of the electromagnetic field along the *z* and *y* axes are subject to a decrease of 12 V/m for the electric field and 2 A/m for the magnetic field.

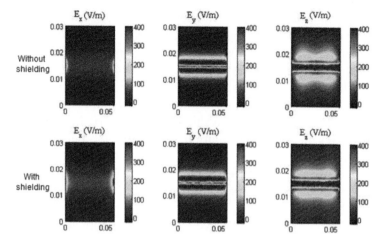

Figure 3.4. *Electric field cartography for f = 3 GHz. For a color version of this figure, see www.iste.co.uk/elhami/embedded1b.zip*

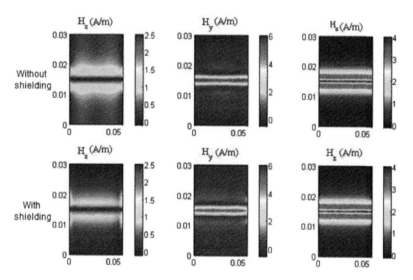

Figure 3.5. *Magnetic field cartography for f = 3 GHz. For a color version of this figure, see www.iste.co.uk/elhami/embedded1b.zip*

For a resonance frequency, for instance, $f_{resonance} = 6.75$ GHz, the electromagnetic field cartographies are subject to a total change accompanied by the appearance of a normal mode of the cavity with index (0, 1, 2) (Figures 3.6 and 3.7). A difference of 50 dBV/m is observed on the longitudinal component of the electric field E_x. For the point with coordinates $x = 5$ mm, $y = 22$ mm and $z = 2$ mm, in the case where the cavity is without shielding, the electric field E_x is equal to 30.85 dBV/m, and if the cavity has shielding, the field is equal to 79.90 dBV/m. For this resonance frequency and knowing that only the field lines of the microstrip line move from the strip toward the ground plane, with the presence of shielding, they also close on the metallic linings of the cavity. This allows the stimulation of the cavity and gives birth to a standing wave representing a normal mode of the cavity [ARC 03, KEH 10].

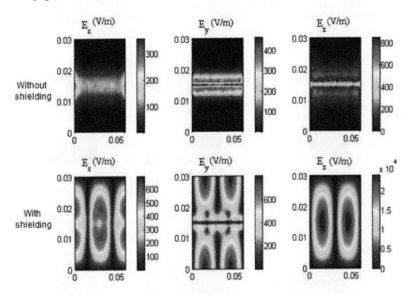

Figure 3.6. *Electric field cartography for $f_{resonance} = 6.75$ GHz. For a color version of this figure, see www.iste.co.uk/elhami/embedded1b.zip*

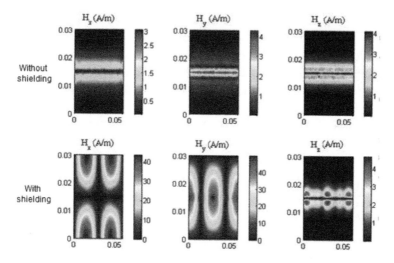

Figure 3.7. *Electric field cartography for $f_{resonance}$ = 6.75 GHz. For a color version of this figure, see www.iste.co.uk/elhami/embedded1b.zip*

3.4. Approximation of the electromagnetic field radiated in the presence of the cavity from the electromagnetic field radiated without cavity

3.4.1. *Principle of the approach*

The designers of hyperfrequency circuits observe malfunctions in their circuits in certain configurations, mainly when circuits are encapsulated within a cavity. For a better understanding of these failures, it is essential to know the electromagnetic environment inside the cavity. In order to measure this field, a measuring probe has to be inserted in the cavity. However, in general, the housings used in radar modules are quite small (a height in the order of 10 mm). This makes the measurement difficult or even impossible to perform. To solve this issue, a new approach is proposed to predict the electromagnetic field radiated by a hyperfrequency module encapsulated in a closed cavity from the field of this same circuit placed in an open cavity (without shielding) [KHE 11].

Figure 3.8 describes the various steps of this approach. After gaining knowledge of the distribution of the electromagnetic field radiated by a hyperfrequency circuit in the configuration without shielding, this field is modeled by a set of equivalent sources called electric dipoles, which radiate the same field as the hyperfrequency circuit [RAM 10, FER 09]. This network of dipoles is then inserted into HFSS. Encapsulation is modeled by changing the limit conditions. The electromagnetic fields radiated by the device placed in a closed cavity are obtained by simulation.

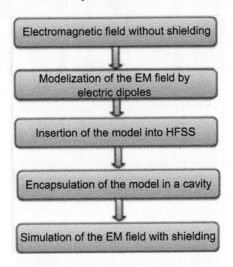

Figure 3.8. *Flowchart of the estimation of the electromagnetic field in closed cavity*

3.4.2. *Radiated emission model*

The radiated emission model [RAM 10] uses a set of electric dipoles to predict the electromagnetic (EM) emissions of the electronic components and systems. The parameters of this model are extracted from measurements of the components of the magnetic field. This radiated emission model has been validated in a variety of microwave devices such as a high-frequency transmission line, a microstrip patch antenna, miniature and "on-a-chip" devices of very small dimensions (space ≈ 10 mm^2) used very frequently in RF-in-package and system-on-a-chip systems.

3.4.2.1. Model topography

The model is represented by a set of elementary electric dipoles placed on an (*xy*)-plane, as shown in Figure 3.9. Each dipole in the network is represented by its position, its orientation (θ) with respect to the *x* axis, its length ($d\ell$) and the current flowing through (I).

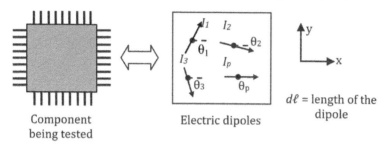

Figure 3.9. *Electric dipole network representing the model*

In this model, the length and position of each dipole are predetermined. However, the orientation of each dipole, its current and the relative effective dielectric constant (ε_{reff}) of the DUT device are retrieved by an automated approach based on an inverse method.

A modeling technique was developed in order to retrieve the three unknown parameters of the model. The mathematical system has to be well defined by three distinct equations:

$$[H_x] = f_{Hx}([\alpha_x],[I],[\theta],[k],fr)$$
$$[H_y] = f_{Hy}([\alpha_y],[I],[\theta],[k],fr)$$
$$[E_z] = f_{Ez}([P_o],[P_d],[I],[\theta],[k],[\varepsilon],fr)$$

[3.9]

where:

– f_r is the modeling frequency;

– $[\alpha_x]$ and $[\alpha_y]$ are constants as a function of the dipolar positions, their lengths and of the observation point;

– $[\varepsilon]$ is the permittivity of the environment given by $[\varepsilon_0 \varepsilon_{reff}]$, in which $\varepsilon_0 = 8.85418e^{-12}$ F/m;

– [k] is the wave number given by [k] = (2π/λ) x [ε_reff]^0.5;

– λ (wavelength) = C/f_r;

– C is the speed of light in air;

– [P_d] is the dipole position vector;

– [P_o] is the position vector of the observation point;

– F_Hx, F_Hy and F_Ez are nonlinear mathematical functions.

3.4.2.2. *Retrieving the model parameters*

The retrieval of the dipolar orientations, currents and the effective relative permittivity is the key part of the construction of the model. The retrieval is performed in two steps. The vectors of the initial parameters are calculated. These parameters are optimized using a nonlinear method based on the Levenberg–Marquardt method.

The model is retrieved on the basis of the following conditions constrained by the user:

– physically $1 < \varepsilon_{reff} < \varepsilon_r$ (from the substrate present in the DUT). The effect of ε_{reff} is negligible on the magnetic field very close to the DUT;

– the dipolar orientations and the currents are first of all retrieved and then the effective relative permittivity is determined.

3.4.2.3. *Obtaining the initial vectors of the model parameters*

The initial vectors of the dipolar orientations $[\theta_{init}]$ are approximated by the inverse method, as shown below:

$$[H_x]_{mx1} = [\alpha_x]_{mxp} \times [I_{init} \cdot \sin(\theta_{init})]_{px1}$$
$$\Rightarrow A_i = [\alpha_x]_{mxp}^{-1} \times [H_x]_{mx1} = [I_{init} \cdot \sin(\theta_{init})]_{px1} \quad [3.10]$$

$$[H_y]_{mx1} = [\alpha_y]_{mxp} \times [I_{init} \cdot \cos(\theta_{init})]_{px1}$$
$$\Rightarrow B_i = [\alpha_y]_{mxp}^{-1} \times [H_y]_{mx1} = [I_{init} \cdot \cos(\theta_{init})]_{px1} \quad [3.11]$$

$$\therefore [\theta_{init}] = \tan^{-1}(A_i/B_i) \in R, \text{ for } i = 1, 2, ..., p \quad [3.12]$$

where:

— m is the number of measurement points;

— p is the number of dipoles used for modeling;

— mathematically, the inverse matrix only exists for square matrices. In the case where $p \neq m$, the pseudoinverses are used to determine the initial orientation vector. The initial current vector $[I_{init}]$ is calculated by retrosubstitution of the values of $[\theta_{init}]$ in [3.10] or [3.11]. The first value of $[\varepsilon_{reff}]$ is the unit vector ($[\varepsilon_{reff\_init}] = 1$) according to the above condition.

3.4.2.4. Optimizing the model parameters

A two-step optimization technique based on the nonlinear Levenberg–Marquardt least-squares method is applied to retrieve the parameters of the model. The system is separated into real and imaginary parts and optimized in such a way that the error functions [3.13] and [3.15] are simultaneously minimized:

$$f_1 = \begin{cases} real\left(H_{x,y}^{meas} - H_{x,y}^{model}\right) \\ imag\left(H_{x,y}^{meas} - H_{x,y}^{model}\right) \end{cases} \quad [3.13]$$

and:

$$H_x^{model} = f_{Hx}\left([\alpha_x], [I_{init}], [\theta_{init}], [k], f_r\right)$$
$$H_y^{model} = f_{Hy}\left([\alpha_y], [I_{init}], [\theta_{init}], [k], f_r\right) \quad [3.14]$$

where:

$$f_2 = |E_z|^{meas} - |E_z|^{model} \quad [3.15]$$

and:

— $[\varepsilon_{init}] = [\varepsilon_0 \varepsilon_{reff\_init}]$;

— $[k_{init}] = (2\pi f r/c)\,[\varepsilon_{reff\_init}]^{0.5}$.

All the definitions of other parameters remain the same as before. The first step of the procedure optimizes the orientations and the currents of single dipoles, without taking into account the effect of the effective dielectric constant ($[\varepsilon_{reff}] = 1$) according to the above condition. The values of I and θ (I_{sol} and θ_{sol}) are retrieved at the first optimization step and are used in the second-step optimization procedure. Such an iterative method makes it possible to retrieve the parameters while preserving their physical meaning: the dipolar orientations are real and their currents are complex with their real and imaginary parts retrieved separately.

3.4.2.5. *Model of the case study*

To validate these methods, we apply them to the microstrip line. The line is simulated in HFSS at a frequency equal to 3 GHz with an open cavity. The simulated fields at a distance of 2 mm above the device are used to build the previously described equivalent model. This model is then inserted into HFSS as shown in Figure 3.10, where the upper surface of the cavity is kept open.

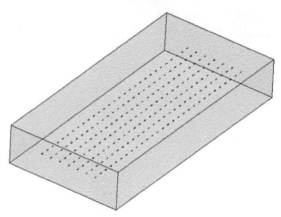

Figure 3.10. *Insertion of the model into HFSS*

3.4.3. *Results and discussion*

The obtained model is simulated and the radiated fields at 2 mm above the microstrip line are compared to those of the actual structure. This encapsulation is performed by a simple change of limit conditions in HFSS.

The cartographies of Figures 3.11–3.14 display the electromagnetic field corresponding to open shielding and closed shielding for the model and the line–cavity system. We observe a good fit between the results with slight differences in certain components of the field. By considering particular sections, we can observe that for the components of the electromagnetic field along the y and z axes, we have small errors between the simulations of the line–cavity structure and the simulation of the model. However, for the components E_x and H_x, the model is not capable of reproducing these two components given the very low value of the field. Thus, we may say that overall, this model makes it possible to predict the effect of the shielding on the cartographies, except for the low-amplitude components which, in other words, have little influence on the structure.

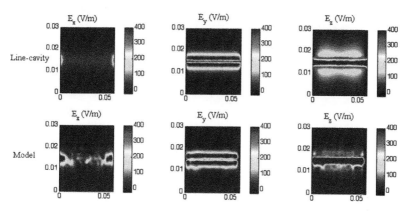

Figure 3.11. *Electric field cartography for f = 3 GHz for open shielding. For a color version of this figure, see www.iste.co.uk/elhami/embedded1b.zip*

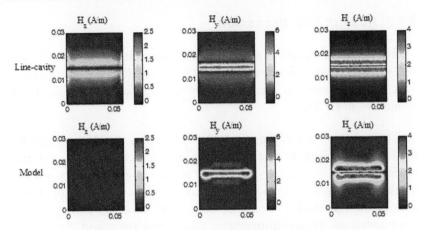

Figure 3.12. *Magnetic field cartography for f = 3 GHz for open shielding. For a color version of this figure, see www.iste.co.uk/elhami/embedded1b.zip*

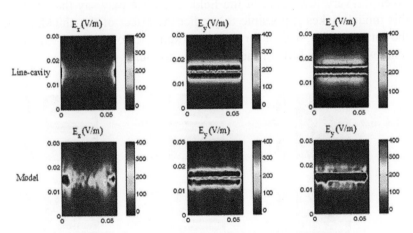

Figure 3.13. *Electric field cartography for f = 3 GHz for closed shielding. For a color version of this figure, see www.iste.co.uk/elhami/embedded1b.zip*

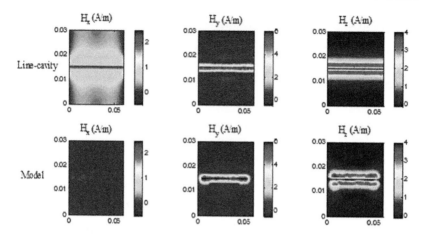

Figure 3.14. *Magnetic field cartography for f = 3 GHz for closed shielding. For a color version of this figure, see www.iste.co.uk/elhami/embedded1b.zip*

We evaluate the capacity of this model to make predictions at other frequencies for which the modes of the cavity are stimulated. These modes impose new distributions of the electromagnetic field that the model is not, at this stage of development, capable of representing (Figure 3.15).

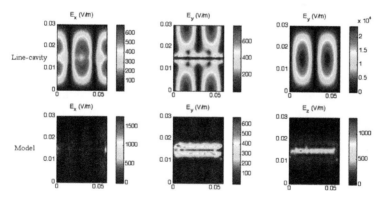

Figure 3.15. *Electric field for electric field cartography for $f_{resonance}$ = 6.75 GHz when the cavity is closed. For a color version of this figure, see www.iste.co.uk/elhami/embedded1b.zip*

Several hypotheses are presented to explain these limitations:

– the dipole model remains a highly narrowband frequency model, which limits its validity on several frequencies. It would be more interesting to reuse this approach with a more temporal, wider-band model;

– knowing that the resonance frequencies are highly localized frequencies, we may assume that the modeling uncertainties can cause the fluctuation and displacement of these frequencies;

– the effective permittivity varies between the two configurations – open and closed cavity – mainly in the case of stimulated modes;

– the model remains a behavioral model which has difficulty answering all the problems such as those discussed in this chapter.

3.5. Conclusion

The prediction of the electromagnetic radiation of a hyperfrequency circuit protected by an electromagnetic cavity requires a sound knowledge of the resonance frequencies and various modes of the cavity stimulated by this circuit. The study of the transmission coefficient for the two configurations (open cavity and closed cavity) makes it possible to determine these frequencies.

In this chapter the use of simulation has shown that the distribution of the electromagnetic field is not altered by the presence of the cavity except close to the cavity walls. However, the simulation of the resonance frequencies showed significant differences between the electromagnetic fields of two configurations (simulation of the S parameters).

The study shows that the components enclosed in a metallic cavity are in a very harsh electromagnetic environment. These results are used to study the effect of this environment on the characteristics of an RF transistor.

Measuring the distribution of the electromagnetic field in a closed cavity remains complicated. Indeed, any induction of the metallic measuring probe is likely to change the distribution of the fields and to modify certain resonances. A method has been developed to obtain this distribution based on the electromagnetic field of an open structure. This procedure consists of modeling the radiated field of the system using a model where a set of electric dipoles is distributed in the space following the two axes. This model is inserted into HFSS and enclosed in a metallic cavity.

This method is validated for all frequencies except for the resonance frequencies of the cavity.

There are two possible extensions of this research. The first extension involves studying the behavior of the equivalent permittivity in two configurations (open shielding and closed shielding). The second extension consists of developing a new cavity model through 3D modeling.

3.6. References

[ARC 03] ARCAMBAL C., Introduction des contraintes de propagation et rayonnement électromagnétiques dans l'étude et la conception d'émetteurs/récepteurs de puissance, PhD thesis, University of Rouen, July 2003.

[DIX 05] DIXON P., "Dampening cavity resonance using absorber material", *Microwave Magazine IEEE*, vol. 6, no. 2, pp 74–84, August 2005.

[FAN 94] FANGYU L., "Perturbed effect of the gravitational wave produced bymicrowave electromagnetic cavity on detecting electromagnetic field", *Chinese Physics Letters*, vol. 11, pp. 321–324, 1994.

[FER 09] FERNANDEZ P., RAMANUJAN A., VIVES Y. *et al.*, "A radiated emission model compatible to a commercial electromagnetic simulation tool", *20th International Symposia on EMC*, Zurich, p. 369, January 2009.

[KHE 10] KHEMIRI S., KADI M., LOUIS A. et al., "Etude de l'effet d'une cavité métallique sur les cartographies du champ électromagnétique d'un circuit hyperfréquence", *15ème Colloque. International. Exposition sur la Compatibilité Electromagnétique*, Limoges, France, April 2010.

[KHE 11] KHEMIRI S., RAMANUJAN A., KADI M. et al., "Estimation of the electromagnetic field radiated by a microwave circuit encapsulated in a rectangular cavity", *IEEE International Symposium on Electromagnetic*, Long Beach, CA, 2011.

[KON 75] KONG J.A., *Theory of Electromagnetic Waves*, John Wiley & Sons, New York, 1975.

[RAM 10] RAMANUJAN A., RIAH Z., LOUIS A. et al., "Modeling the electromagnetic radiations of passive microwave components using a near-field scanning method", *IEEE Transactions on Electromagnetic Compatibility*, vol. 52, no. 4, pp. 254–261, 2010.

4

Metrology of Static and Dynamic Displacements and Deformations Using Full-Field Techniques

This chapter presents an experimental investigation, by full-field optical metrology, of static and dynamic deformations in electronic equipment and components. The different metrology techniques employed produce results that are complementary in terms of temporal and spatial resolution, as well as measurement sensitivity. This facilitates the study of specific phenomena related to thermo-mechanical and vibration behavior of the electronic devices under test. The experimental results can be useful in the development and validation of numerical models for the reliability analysis of mechatronic systems.

4.1. Introduction

Optical measurement techniques have the distinct advantage that no physical contact with the test surface is required. This allows us to study small amplitude mechanical and thermal deformations. An important aspect of optical metrology is the absence of transducers, meaning that there is no thermal or mechanical influence from the measurement process. In the general context of optical measurement techniques, "full-field" techniques are represented, in our case, by:

– interferometric techniques – holographic interferometry (HI) and speckle interferometry (SI);

Chapter written by Ioana NISTEA and Dan BORZA.

– non-interferometric techniques – projection Moiré (PM), structured light (SL) techniques or digital image correlation (DIC).

These techniques have several specific characteristics:

– the raw data are acquired as images, simultaneously for all visible points of the test object;

– the images are obtained by illuminating (Figure 4.1) all the visible points $P(x,y,z)$ of the test object by using a coherent point source, more specifically a laser beam diverging from point $S(x_0,y_0,z_0)$;

– the measurement results can alternatively be presented as black and white (B&W) images, where the gray level associated with each pixel represents the value of the displacement at that point.

The images are acquired by a detector (holographic film in HI, charge coupled device (CCD) camera in SI, etc.), identified at each point as $H_k(x_k,y_k,z_k)$. The distance between $S(x_0,y_0,z_0)$ and $P(x,y,z)$ is R_0, while the distance between $P(x,y,z)$ and $H_k(x_k,y_k,z_k)$ is R_k. Between two subsequent acquisitions, a given point $P(x,y,z)$ is displaced to $P'(x+u,y+v,z+w)$.

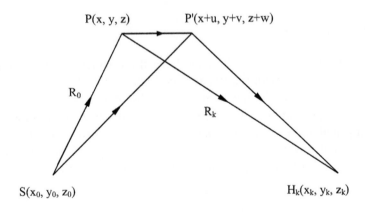

Figure 4.1. *Point P displacement to P'*

For all the techniques mentioned above, the fringe pattern $I(x,y)$ obtained by applying specific algorithms to the acquired images is described by:

$$I(x,y) = I_0(x,y) \cdot F[\Delta\phi(x,y)] \qquad [4.1]$$

In equation [4.1], $I_0(x,y)$ is the object image, while F is the so-called fringe function. $\Delta\phi(x,y)$ represents the optical phase difference, whose value depends on the displacement of each point $P(x,y,z)$ and on the interferometric setup (out-of-plane, in-plane or shearography).

More specifically, in HI and SI we have:

$$\Delta\phi = \frac{2\pi}{\lambda}(A_k u + B_k v + C_k w) \qquad [4.2]$$

The coefficients from equation [4.2] are given by:

$$A_k = A_k(x,y,z) = \frac{x - x_0}{R_0} + \frac{x - x_k}{R_k}$$

$$B_k = B_k(x,y,z) = \frac{y - y_0}{R_0} + \frac{y - y_k}{R_k}$$

$$C_k = C_k(x,y,z) = \frac{z - z_0}{R_0} + \frac{z - z_k}{R_k} \qquad [4.3]$$

with the coefficients $R_0(x,y,z,x_0,y_0,z_0)$ and $R_k(x,y,z,x_k,y_k,z_k)$ representing, respectively:

$$R_0 = \sqrt{(x-x_0)^2 + (y-y_0)^2 + (z-z_0)^2} \qquad [4.4]$$

and

$$R_k = \sqrt{(x-x_k)^2 + (y-y_k)^2 + (z-z_k)^2} \qquad [4.5]$$

4.2. Speckle interferometry

The main components of an SI setup are:

– coherent source (laser, mirrors and lenses) for illuminating the object;

– interferometer;

– camera, allowing for the recording of the optical fields at the output of the interferometer;

– acquisition and control software.

Depending on the configuration of the interferometer, this type of setup can be used to measure out-of-plane or in-plane displacement fields, as well as the first derivatives of the out-of-plane displacement. In the context of the analysis of failures of embedded mechatronic systems (AUDACE) project, only the first type of SI measurement setup (out-of-plane sensitivity) was employed and therefore will be described. This configuration is also known as "electron holography".

4.2.1. *Principles of displacement field metrology by speckle interferometry*

Electron holography is the most widely known SI configuration. Its most significant characteristics, in the wide spectrum of SI techniques, are:

– there is one single beam illuminating the object and the wave scattered from the surface of the object contains speckles. It passes through the camera lens to form a plane image on the CCD sensor;

– a uniform reference wave, having the same direction of propagation as the object wave, is introduced between the camera lens and the CCD. It combines with the object wave, creating an interference field which is integrated by the camera during the exposition of each frame.

This is, therefore, an "in-line" holographic setup, as shown in Figure 4.2. In this case, we can apply equation [4.1] and the following to describe the recording of the hologram. There are two notable differences between this setup and HI: the object wave contains speckle, due to the fact that it passes through an optical system, and the primary fringe spacing is significantly larger than that of HI, to be compatible with the (relatively low) resolution of the CCD sensor. A third difference is, of course, the interpretation of the primary interferogram formed on the CCD sensor; the object wave restitution by diffraction is replaced by numerical algorithms for extracting phase information directly from the primary interferograms (holograms).

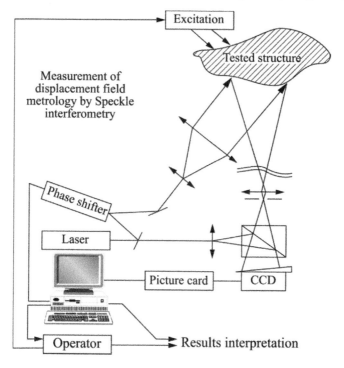

Figure 4.2. *Schematic diagram of an electronic holography setup*

If the illumination and the observation directions are approximately collinear, the system's sensitivity vector will have the same orientation. In such a case, if the object surface is perpendicular to the three directions (z), the system will have maximum sensitivity to the out-of-plane displacement (w).

In order to implement various modulation and temporal phase stepping algorithms, devices capable of generating a known variation in the optical path are placed on the paths of the reference beam and the illumination beam. The simplest solution is to use adjustable mirrors that can be displaced by means of lead zirconate titanate (PZT) actuators.

The most straightforward measurement method is time averaging, used in the case of vibrating objects. In its simplest implementation, this method facilitates the measurement of vibration amplitude fields in the case of stationary vibration. It will serve as a starting point for further analysis and development.

4.2.1.1. *Time-average speckle interferometry*

The complex amplitude of the object wave incident on a point (x,y) of the CCD detector can be described by:

$$\bar{U}_O(x,y) = O(x,y) e^{j\phi_O(x,y,t)} \tag{4.6}$$

If the object is vibrating harmonically at a frequency that is significantly higher than the acquisition rate of the camera, then:

$$\phi_O(x,y,t) = \phi_{O0} + \frac{4\pi}{\lambda} \cos \omega t \tag{4.7}$$

The object wave interferes with a reference wave, written as:

$$\bar{U}_R(x,y) = R(x,y) e^{j\phi_R(x,y)} \tag{4.8}$$

The camera averages, with an exposure time of $T_0 >> \dfrac{2\pi}{\omega}$, the intensity corresponding to the interference field $\bar{U} = \bar{U}_R + \bar{U}_O$, resulting in:

$$I = (O^2 + R^2) + 2\sqrt{O^2 R^2} \cdot \cos(\phi_{OO} - \phi_R) \times J_0(\phi_v) \qquad [4.9]$$

In equation [4.9], the notation

$$\phi_v(x, y) = \dfrac{4\pi}{\lambda} d(x, y) \qquad [4.10]$$

was used to designate the phase variation due to the vibration amplitude $d(x,y)$. $J_0(\phi_v)$ represents the zero-order Bessel function of the first kind, having ϕ_v as its argument.

By applying a temporal phase step of $\dfrac{\pi}{2}$ between any two consecutive images, the mathematical expressions of any group of four acquired images can be obtained from equation [4.9]. The following notations will be used: $A = O^2 + R^2$, $B = 2\sqrt{O^2 R^2}$ and $\phi_{O-R}(x, y) = \phi_{OO} - \phi_R$.

It should be noted that the phase distribution $\phi_{O-R}(x, y)$ is random and that the expressions of the intensities, A and B, contain speckle noise. The noise appears both as an additive and a multiplicative term. The expressions derived for the four primary interferograms are:

$$I_i = A + B \cdot \cos\left[\phi_{O-R} + (i-1)\dfrac{\pi}{2}\right] \cdot J_0(\phi_v) \; ; i = 1, 2, 3, 4 \qquad [4.11]$$

The time-averaged hologram can be calculated by eliminating A and ϕ_{O-R} from these expressions:

$$I_{IT} = \sqrt{(I_1 - I_3)^2 + (I_4 - I_2)^2} = 2B|J_0(\phi_v)| \quad [4.12]$$

Thus, the fringe patterns obtained (Figure 4.3, vibration mode of a ventilator casing, at 66.8 Hz and a vibration mode for a motor controller printed circuit board (PCB) fitted in its enclosure, at 243 Hz) contain the necessary data for calculating the amplitude fields of the respective modes.

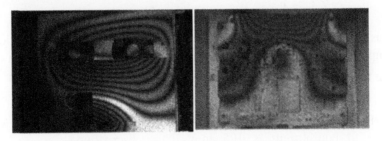

Figure 4.3. *Time-averaged electronic holograms*

The consecutive images described by equation [4.10] are updated continuously. The natural frequencies can be determined with high precision by identifying, around every resonance, the excitation frequency that generates a maximum number of iso-amplitude fringes.

The quantitative interpretation of the result given by equation [4.11] needs to be backed up by additional information on the relative vibration phases of the various points on the surface. As a general rule, when crossing a nodal line, the vibration phases are inverted.

The difficulties related to automatic data analysis by monitoring the fringes are due to speckle noise (the B term in equation [4.12]) being higher than in HI; to the decreasing fringe contrast caused by the gradual decay of the bright fringes, given by the extremes of $J_0(\phi_v)$, toward higher amplitudes of vibration; and to high density of fringes in certain areas of the object, resulting in undersampled spatial frequencies.

These challenges require finding solutions through improvements made to the SI measurement system.

To help with interpreting the vibration fringe patterns of the tested electronic components that will be presented in subsequent chapters, a typical Bessel fringe pattern obtained for a rectangular plate excited on one of its vibration modes is presented in Figure 4.4, as well as the respective three-dimensional (3D) surface displacement map.

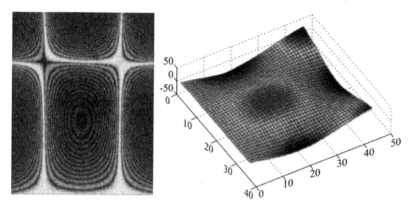

Figure 4.4. *Time-averaged fringe map of a vibration mode and its interpretation presented as a surface displacement map*

4.2.1.2. Metrology of static deformations of thermal or mechanical origin

The four frames recorded with the object in its initial state are:

$$I_i = A + B \cdot \cos\left[\phi_{O-R} + (i-1)\frac{\pi}{2}\right]; i = 1, 2, 3, 4; \qquad [4.13]$$

After the object has deformed, the object wave's optical phase displays, just like in HI, a variation $\Delta\phi_{OBJ}$:

$$\Delta\phi_{OBJ} = \vec{K} \cdot \vec{D} \qquad [4.14]$$

with $\vec{K}$ being the sensitivity vector and $\vec{D}$ the displacement of a point on the surface of an object, as defined in section 4.1. The four frames acquired after the deformation are:

$$J_i = A + B \cdot \cos\left[\phi_{O-R} + \Delta\phi_{OBJ} + (i-1)\frac{\pi}{2}\right]; i = 1, 2, 3, 4 \quad [4.15]$$

There are two approaches to calculate $\Delta\phi_{OBJ}$:

– generating an intensity fringe pattern;

– generating a phase fringe pattern.

In the first approach, without going into the detail of the calculation, it can be shown that for extracting $\Delta\phi_{OBJ} = \vec{K} \cdot \vec{D}$ the formula given in equation [4.16] can be applied:

$$I_{2E} = \sqrt{(I_1 - I_3 + J_1 - J_3)^2 + (I_4 - I_2 + J_4 - J_2)^2} = 8B \cdot \left|\cos\frac{\Delta\phi_{OBJ}}{2}\right| \quad [4.16]$$

The result is equivalent to the reconstructed image in HI. The contrast of the fringe pattern is higher than the one calculated for vibration measurement, but the amplitude noise, represented by the term B, is still present.

In order to obtain more directly the phase distribution by creating a phase map, $\Delta\phi_{OBJ}|_{\mod 2\pi}$ (the wrapped phase, modulo 2π) can be calculated directly by:

$$I_{PH} = \text{arctg}\frac{(I_1 - I_3)(J_4 - J_2) - (I_4 - I_2)(J_1 - J_3)}{(I_1 - I_3)(J_1 - J_3) + (I_4 - I_2)(J_4 - J_2)} = \Delta\phi_{OBJ}|_{\mod 2\pi} \quad [4.17]$$

This second approach is more useful, as it eliminates phase uncertainties, which are inherent when using the cosine function: $\left|\cos\frac{\Delta\phi_{OBJ}}{2}\right|$. For the often encountered case of out-of-plane displacements d(x, y) with the object illumination and the observation lying along the same direction, we have:

$$I_{PH} = \Delta\phi_{OBJ}\big|_{\mathrm{mod}\,2\pi} = \frac{4\pi}{\lambda} \cdot d(x,y)$$

[4.18]

The image shown in Figure 4.5 presents the phase distribution $\Delta\phi_{OBJ}\big|_{\mathrm{mod}\,2\pi}$ for a rectangular plate clamped on its sides and excited by an out-of-plane force. The test is performed as part of a research project concerning a new method of identifying elastic parameters.

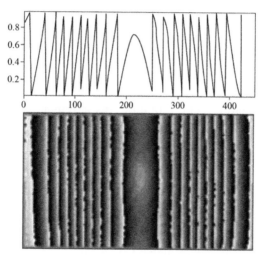

Figure 4.5. *Wrapped phase hologram of a rectangular plate in bending. Above, the profile along a horizontal line in the fringe map*

In order to eliminate the $-\pi$ to π phase jumps, a phase unwrapping procedure has to be run. This topic, however, is beyond the scope of this book.

The images presented in Figure 4.6 illustrate: (1) to the left, thermal deformations of a PCB around the perimeter of a microprocessor, as wrapped fringe map and (2) to the right, the phase distribution (and consequently, the out-of-plane displacement) of the microprocessor alone, as calculated by applying an unwrapping algorithm.

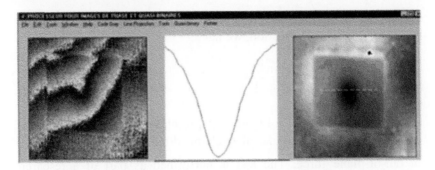

Figure 4.6. *Out-of-plane displacement map of a microprocessor relative to the overall displacement of the PCB*

4.2.2. Description of the speckle interferometry measurement setup

The SI setup is designed for the non-contact measurement of thermal deformation and vibration displacements in mechatronic assemblies.

The main components of the SI system are:

– an anti-vibration table on air suspension;

– a continuous frequency-doubled, single-longitudinal-mode Nd:YAG laser;

– a set of optical, electronic and mechanical components, including two adjustable mirrors with piezoelectric actuators;

– a computer with a set of four ISA acquisition cards;

– two monitors, cables, connectors and various mechanical supports;

– a B&W CCD detector acquiring at 25 frames/s (480 × 512 pixels) with several camera lenses;

– control software for the acquisition cards and the electronic devices;

– specifically developed software to analyze the raw data and present the results;

– various auxiliary electronic devices (oscilloscopes, signal generators, amplifiers, power supplies, etc.).

The system facilitates the measurement of static and slowly varying displacement fields using modulo 2π phase imaging, as well as stationary vibrations at the natural frequencies of the tested objects, using time averaging.

The images in Figure 4.7 show several of the main components.

Figure 4.7. *Main components of the speckle interferometry system*

In order to facilitate the setup's development and to test its effectiveness in characterizing reliability defects, it is necessary to:

– characterize the deformations of a mechatronic system for which there exists accurate reliability data (motor controller card);

– find connections between the test results and the reliability defects observed;

– improve the efficiency of the setup, if necessary.

4.2.3. *Examples of static displacement field measurements*

In order to develop a procedure for the optical metrology of out-of-plane deformations in static and quasi-static regimes, we study:

– the effect of fastening the embedded electronic boards onto their enclosures with screws and the effect of the order in which the screws are fastened on either the PCB's or the enclosure's deformation;

– two populated motor controller cards (measurement of thermo-mechanical deformations produced by convection heating);

– measurement of deformations with heat dissipation of a casing with heating elements (resistors) that simulate power transistors.

SI measures out-of-plane deformations occurring between the two states: an initial state and a final state.

To obtain an agreement between the maximum deformation measurable by SI and the deformation obtained by tightening the eight screws to nominal torque, as well as having stable and repeatable initial conditions, all the screws are initially tightened to half of the nominal torque, that is, 100 mNm. This torque value is applied by means of a torque screwdriver, as specified by the manufacturer.

4.2.3.1. The effect of fastening the embedded electronic boards onto their enclosures with screws and the effect of the order in which the screws are fastened on either the PCB's or the enclosure's deformation

Measurements are performed for each of the eight screws shown in Figure 4.8, by applying additional tightening. Real-time monitoring of the resulting deformations allows establishing the values of additional clamping torques.

Figure 4.8. *The eight mounting screws of the PCB*

These values are different from one screw to the other.

To prevent displacements caused by the deformation or rotation of the enclosure during the tightening of the screws, supplementary support elements are added so as to block any movement in the horizontal direction and to fasten the assembly to the optical table (Figure 4.9).

Figure 4.9. *Enforced enclosure for the electronic boards equipped with power transistors or simulators*

Figure 4.10. *To the left, deformations produced by tightening screw 3 × 60 mNm. To the right, deformations produced by tightening screw 5 × 50 mNm*

The testing procedure, with an initial preload of 100 mNm for all screws, is chosen after several attempts at tightening less gradually, straight to the value of 200 mNm, which produced deformations that exceed the measurable limits for this technique, as shown in Figure 4.10.

We find that the eight screws can be separated into two groups. The first group, the four screws in the "corners" (screws 1, 4, 5 and 8 in Figure 4.8), is characterized by low values of deformation. The second group (the "inner" screws 2, 3, 6 and 7) is characterized by much larger deformations.

The four interferograms shown in Figure 4.11 represent the out-of-plane deformation of the electronic board the four figures of interferograms (screws 1, 4, 5 and 8). The interferograms in Figure 4.11 are positioned according to the respective screws (see Figure 4.8). The value of the applied torque is marked below each interferogram.

Figure 4.11. *Deformation of the electronic board due to tightening of the individual screws 1, 4, 5 and 8*

Figure 4.12. *Deformation of the electronic board due to tightening of the individual screws 2, 3, 6 and 7*

For these screws, under the conditions of the test series, the maximum out-of-plane displacement value varies between 5 nm/mNm (for screw 5) and 42.5 nm/mNm (for screw 1).

The interferograms obtained for the second group of screws (2, 3, 6 and 7) are shown in Figure 4.12.

For these four screws, in the conditions of the test series, the maximum out-of-plane displacement value varies between 106 nm/mNm (for screw 6) and 297 nm/mNm (for screw 7) – about an order of magnitude higher than the values obtained for the first group of corner screws.

4.2.3.2. Two populated motor controller cards (measurement of thermo-mechanical deformations produced by convection heating)

Parts of the equipment used for this series of tests are already in use at the Photomechanics laboratory (an adjustable SI measurement system, a heating device built in the laboratory, various holders and supports for the tested components, computer software to perform the data processing and thermocouples).

Within the context of the research program, this equipment is further developed by the purchase and integration into the test setup of a new PC with image and data acquisition cards, an infrared camera, adjustable brackets, devices required for high-speed interferometry – a non-resonant fast phase modulator with its amplifier and a continuous wave single-longitudinal-mode frequency-doubled YAG laser, polarization – maintaining single mode optical fiber with its alignment devices, and the optical elements (dual lens system and holders) required to achieve the interferometer. These upgrades of the SI measurement system take into account the requirements related to measurement trials scheduled for the various test objects.

To illustrate the suitability of this technique for measuring the deformations of the controller card, Figure 4.13 shows the deformations of the PCB (made of FR4) relative to an initial time $t0$ occurring at two instants, $t1$ and $t2 = t1 + 1$ min. After measuring the initial time, stress is applied to the PCB by slackening the central screw. On the left, we see the wrapped phase and on the right, we see the unwrapped phase and the values along the selected horizontal line.

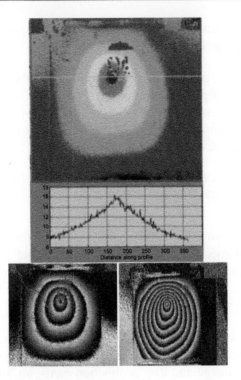

Figure 4.13. *PCB deformation following the slackening of the central screw*

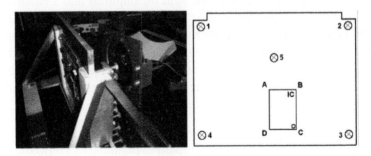

Figure 4.14. *On the left, the adjustable support of the assembled casing or the bare PCB; on the right, the location of the fastening screws and the controller's microprocessor*

The PCB support/housing is mounted vertically on the optical table, with an adjustable support comprising two very rigid steel bars, in order to avoid any parasitic deformation. The two bars are fastened to a base plate that is bolted to the optical table. The support, as well as the location of the five screws and the microprocessor, is illustrated in Figure 4.14.

A K-type thermocouple placed in the central area of the card is used to estimate the temperature change applied locally. Heating is carried out by convection, using a hot air gun.

The image in Figure 4.15 shows the front panel of the software used for temperature control and acquisition triggering. The two graphs represent temperatures measured by the two thermocouples located on either side of the controller card. Numerical values are measured twice per second and written to an American standard code for information interchange (ASCII) file. The pulses used for triggering the holographic recordings can be seen in the upper part of the plot. These pulses can be set to run either at regular time intervals, for example, every 5 s, or at regular temperature intervals for either of the monitored temperatures, e.g. for

$$T(n) = T_{max} - n \cdot 0.2°C \text{, with n = 1, 2, 3 ...} \qquad [4.19]$$

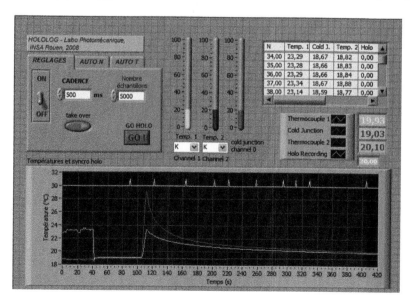

Figure 4.15. *Front panel of the control software used for temperature monitoring and synchronization with the data acquisition*

From this series of recordings, the interferogram representing the out-of-plane deformation field relative to the temperature change T(n1) − T(n2) can be calculated between any two of these instants. Some interferograms are not usable, either because of speckle decorrelation or as a result of small movements or temperature variations of the air that affect the stability of the object or the interferometry system.

Figure 4.16 shows several wrapped phase maps corresponding to deformations of the electronic card assembly under thermal stress produced by convection heating.

Figure 4.16. *Deformation produced by thermal stress of the populated PCB at various instants*

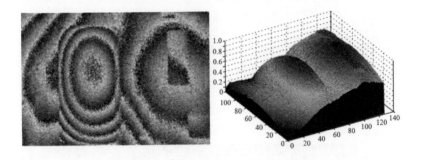

Figure 4.17. *Discontinuities of the deformations produced by thermal stress between the active electronic components and PCB*

Their most significant feature is the discontinuity in the fringe system (Figure 4.17), in most regions where active electronic components are situated, due to the differences in mechanical and thermal properties between components and PCB or copper layers.

A detailed analysis [BOR 12a] showed that these discontinuities result in mechanical stress in the pins of the integrated circuit and in the solder joints on the PCB. These mechanical stresses are aggravated by the presence of the central screw, which amplifies the local curvature of the controller card in the proximity of the microprocessor's solder joints during thermal deformations.

4.2.3.3. *Measurement of deformations with heat dissipation of a casing with heating elements (resistors) that simulate power transistors*

The heating elements which simulate power transistors are mounted onto the support, as shown in Figure 4.9. We use a stabilized power supply that can provide an output current of 2 A at a voltage of 10 V. The real-time monitoring by SI of out-of-plane deformations during heating shows that maintaining a current of 1.5–2 A for a period that exceeds 10–15 s produces increasing deformations in the heating pads and in the support as a whole, and that the fringe system becomes too dense to allow quantitative measurements (Figure 4.18), particularly around the heating pads and the power resistors.

The unwrapped interferogram that allows estimating the displacement field is obtained for a current of 1.7 A, with an output voltage of 7.3 V, applied for 10 s. The deformation map obtained by unwrapping the phase is shown in Figure 4.18.

Figure 4.18. *Speckle interferograms during the gradual heating of the substrate by heating elements (resistors) simulating power transistors*

In addition to an overall deformation of the support–substrate assembly, there is a local maximum in the deformation of the heating elements that simulate power transistors. The profile of this deformation is shown in Figure 4.20.

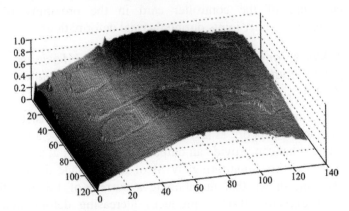

Figure 4.19. *Out-of-plane deformation field which corresponds to the second interferogram presented in Figure 4.18*

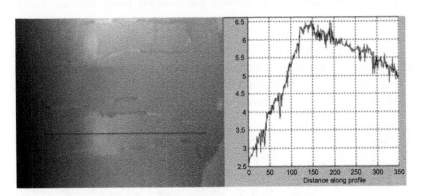

Figure 4.20. *Values of the out-of-plane displacement along the selected line – corresponding to the deformation field shown in Figure 4.18*

4.2.4. *Examples of measurements of vibration displacements fields*

The original plan called for the experimental analysis of a controller assembled in its casing. The interest stirred by the first results led us to perform several series of additional tests:

– bare PCB fastened to a rectangular plate with five screws;

– bare PCB fastened to a rectangular plate with four screws;

– assembled controller fastened to a rectangular plate and mounted in its casing with five screws;

– assembled controller fastened to a rectangular plate and mounted in its casing with four screws;

– specific tests for determining the modes and natural frequencies of the microcontroller and its pins;

– a simplified finite element (FE) model developed in MATLAB® for the first verification of the consistency between the experimental results.

The tests are carried out by out-of-plane time-average SI. The laser used: Nd-YAG, single-longitudinal-mode, 100 mW. The scanning for modes and natural frequencies is performed by exciting the test object into forced vibrations, by means of an amplified signal generator. The signal is monitored using an oscilloscope and is applied to the object (controller mounted in its case) either by direct contact with a piezoelectric vibrator or by sound pressure, using a loudspeaker.

One of the supports providing boundary conditions during the different series of tests for the determination of natural frequencies and modes is shown in Figure 4.21.

Figure 4.21. *Support for studying the vibrations of the controller assembly*

It is worth noting that the influence of the supports in the metrology of static deformations and vibration by SI is extremely important. The measurement sensitivity of this technique is highly elevated; therefore it not only shows the deformations and vibrations of the objects subjected to various stresses, but also those of the supports themselves. Consequently, the quantitative interpretation of the fringe patterns appearing on the test object's surface becomes overly complex. For this reason, the best compromise needs to be found between substrates with the maximum possible stiffness and supports that as closely as possible simulate those used in real-life operation of the tested objects.

Some of the results obtained during the scheduled tests are presented further, in section 4.2.4.

The main operational deflection shapes, very similar to the mode shapes, and their frequencies, are presented in the fringe patterns included in Figure 4.22, for the case of a bare PCB. Their quantitative interpretation is made using the mathematical relations and explanations given in section 4.1 (specifically Figure 4.1 and equation [4.12]).

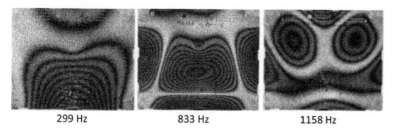

| 299 Hz | 833 Hz | 1158 Hz |

Figure 4.22. *Several operational deflection shapes and their corresponding natural frequencies for the bare PCB fastened with five screws*

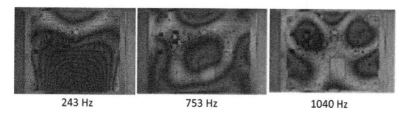

| 243 Hz | 753 Hz | 1040 Hz |

Figure 4.23. *Several vibration modes and their natural frequencies for the populated PCB, fastened with five screws*

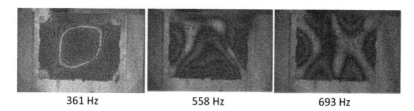

| 361 Hz | 558 Hz | 693 Hz |

Figure 4.24. *Several vibration modes and their natural frequencies for the populated PCB, fastened with four screws*

For the populated PCB fastened by five screws, several vibration modes are illustrated in Figure 4.23.

For the populated PCB fastened only with four screws at the corners, several vibration modes are illustrated in Figure 4.24.

We find that the presence of the fifth screw (in the central position) causes concentrations of displacement around that area, and therefore high values of the spatial derivatives which have a direct relation to the mechanical stresses. To support this statement, we present in Figure 4.25 the vibration amplitude map w (x, y) related to the fringe pattern shown for f = 544 Hz, together with its spatial derivatives $\frac{\partial w(x,y)}{\partial x}$ and $\frac{\partial w(x,y)}{\partial y}$.

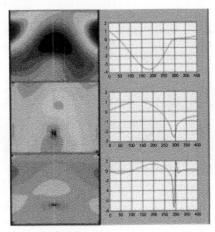

Figure 4.25. w(x, y), $\frac{\partial w(x,y)}{\partial x}$ and $\frac{\partial w(x,y)}{\partial y}$ for a frequency of 544 Hz

Figure 4.26. *Resin encapsulated casing equipped with a controller card, under vibration testing*

The clamping and the vibration excitation devices for areas in encapsulated casing equipped with a controller card are illustrated in Figure 4.26.

Some of the vibration modes are presented in Figure 4.27.

Another series of measurements shows the resonance modes of the integrated circuit alone, as shown in Figure 4.28.

This type of vibration is liable to produce fatigue failures at the solder joints of the integrated circuit. These pins may have, in addition, their own vibration modes, two of which are shown in Figure 4.29.

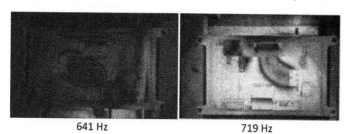

641 Hz   719 Hz

Figure 4.27. *Two vibration modes of the epoxy encapsulated casing/controller card assembly*

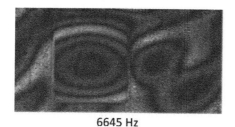

6645 Hz

Figure 4.28. *Modes of vibration of the integrated circuit alone*

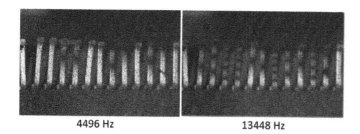

4496 Hz   13448 Hz

Figure 4.29. *Two modes of vibration of the IC pins*

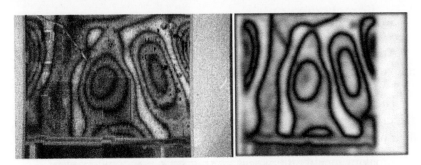

Figure 4.30. *Interferograms obtained by the conventional time-averaging method (on the left) and the high spatial resolution time-average speckle interferometry (on the right)*

In addition to the flexibility allowed by the zoom setting of the camera lens, the spatial resolution of the measurement can be increased by applying high spatial resolution (subpixel) acquisition and processing techniques [BOR 05], and this despite the large speckle noise that characterizes the images obtained by SI. This increase in spatial resolution can be observed by comparing the two interferograms shown in Figure 4.30. The left-hand side shows an interferogram obtained by the conventional time-averaging method, while the right-hand side shows that obtained by the high spatial resolution time-average SI.

The high spatial resolution of this technique led to the implementation of a new algorithm for obtaining the unwrapped phase map from a time-average interferogram. Obtaining the unwrapped phase map is based on the mathematical inversion of the zero-order Bessel function of the first kind on its monotonous intervals [BOR 06].

Full field techniques not only allow high spatial resolution, but also a high temporal resolution, as will be shown in the next section.

4.2.5. *Examples of dynamic measurements*

The high temporal resolution measurement system is built in its entirety at National Institute of Applied Sciences (INSA) Rouen by the authors of this chapter, as part of the research conducted in the field of vibro-acoustics [MOR 08] and a PhD thesis [NIS 10]. This work leads to the completion of a low-cost metrology system, combining spatial resolution with temporal resolution measurement resolution. Several papers [NIS 13] and [BOR 12] present the principles of the invention, their implementation and data processing algorithms used.

The metrology system is tested as part of the AUDACE research agreement. The dynamic measurements concern a subset of a radio frequency (RF) power transistor subjected to 0.5 ms long pulses, during which the transistor undergoes strong thermal stress, followed by slow cooling for several tens of milliseconds.

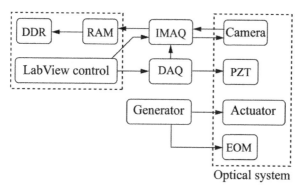

Figure 4.31. *Temporal SI system for dynamic deformation metrology*

The size of the measured surface is 0.2 mm × 0.2 mm. Full series of 5,000 acquired specklegrams allows for the temporal analysis of several cycles of heating and cooling. The specklegrams used for calculating the two states of deformation shown in Figure 4.32 are acquired in 100 × 100 pixel format.

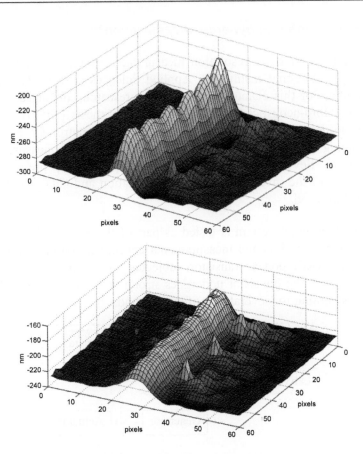

Figure 4.32. *Two deformation states, 0.7 ms apart, of the HF power transistor*

4.3. Moiré projection

Moiré projection techniques are among the first techniques used in optical metrology of surface topography and deformations, and that is why they are found today in many different forms. In the most general of terms, the common trait in all of these techniques is the projection onto the object surface of one or several grids of straight lines (usually having a square or a sinusoidal profile) and the use of the acquired images to obtain the object's coordinates or the displacement field undergone by the object between the two states.

4.3.1. Measurement principles of Moiré projection for displacement fields

By projecting a pattern of linear fringes having a sinusoidal profile (Figure 4.33) onto the object, the shape of the fringes is changed by the relief of the object.

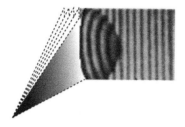

Figure 4.33. *Illuminating the object with a pattern of linear fringes*

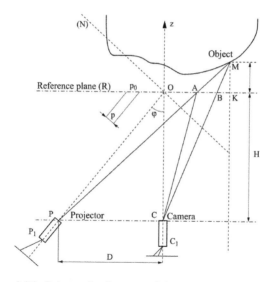

Figure 4.34. *Schematic diagram of the moiré projection setup*

The image of the object is acquired using a camera whose axis CC1 intersects (Figure 4.34) the PP1 axis of the projector in a point O (cross-axis setup) at an angle φ. The virtual reference plane (R) is perpendicular to the axis of the camera, and (N) represents the plane through O that is perpendicular to the axis of the projector.

Among the main equations, we have:

$$z_M = \frac{H \Delta x}{D - \Delta x} \quad [4.20]$$

with:

$$\Delta x = x_B - x_A \quad [4.21]$$

Taking into consideration equation [4.20], it can be demonstrated that the object's full phase is given by:

$$\phi_{OB} = 2\pi f_0 x + \phi_0 + 2\pi f_0 \frac{D z_M}{H + z_M} \quad [4.22]$$

4.3.2. Description of the Moiré projection measurement setup

Moiré projection measurement setups may use either video projection (using commercial projectors, possibly with modified projection lens) of MATLAB®-generated images or projection of fringes obtained from an interferometer (Mach-Zender) with a continuous-wave HeNe laser and spatial phase shifting (using piezoelectric actuators). The phase stepped fringe projection is implemented, in this case, with a commercial video projector. The laboratory has several video projectors, in order to meet specific requirements related to the dimensions of the tested objects.

The following components are also part of the setup:

– a CCD camera, acquiring at 30 frames/s and a compatible image acquisition card, or alternatively, a complementary metal oxide semiconductor (CMOS) Firewire digital camera (without image acquisition card) + a computer;

– dedicated software, developed in MATLAB® by the authors;

– synchronization elements, cables, connectors and various mechanical holders;

– auxiliary electronic devices.

A measurement system of this type, built in our laboratory, is partially shown in Figure 4.35.

Figure 4.35. *A part of the Moiré projection system*

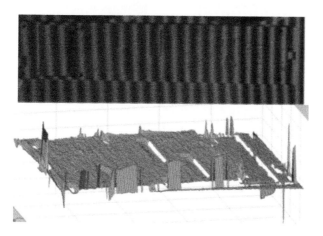

Figure 4.36. *Above, one of the images acquired during measurements; below, the shape differences between the bare aluminum substrate and the metal brazed substrate*

4.3.3. *Examples of displacement field metrology by Moiré projection*

Measurement by Moiré projection (Figure 4.36) shows the differences between a bare aluminum substrate and a metal brazed marble substrate.

4.4. Structured light projection

This technique consists of the projection by means of a video projector or a dedicated analog projector, of a set of light patterns. Most often, these are binary parallel lines (black and white). Figure 4.37 (left-hand side) shows, one above the other, a small part of each frame belonging to such series. The images are averaged so that the succession of values "0" and "1" that make up each binary word [Bit 1 Bit 2 Bit 3 Bit N ...] of each column in the set of images represents in reversed binary code (Gray code) numbers in ascending order: 0, 1, 2, 3, etc. (Figure 4.37, right-hand side).

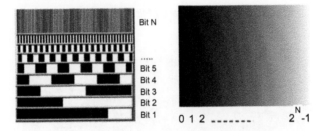

Figure 4.37. *Structured light projection*

Figure 4.38. *Measurement principles of structured light projection*

4.4.1. *Principles of structured light projection for measuring surface topography*

During the projection of images onto the scene (the object) to be scanned, the projected patterns are distorted by the shape of the object (Figure 4.38). After the acquisition, a binary number is obtained for each pixel of the image, which is converted into a decimal number.

Following this principle, the coding is performed along the horizontal direction of the space containing the projector and the object, while the vertical plane projected by the projector (Figure 4.39) and containing the point M is a known quantity.

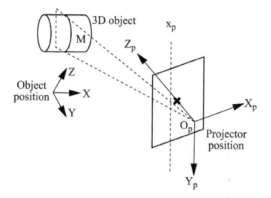

Figure 4.39. *Coding of the projector-object space*

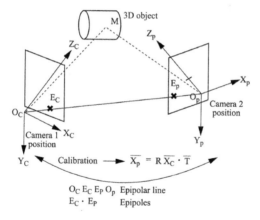

O_C E_C E_P O_P Epipolar line
$E_C \cdot E_P$ Epipoles

Figure 4.40. *Triangulation allowing us to find the position of M*

The full calibration of the measuring system, including the camera and the projector with their respective positions, can be performed by a similar procedure. Determining the 3D position of each point M is achieved by triangulation, by intersecting the vertical plane (connecting the center of the projector, O_P, with M) with the segment $O_C M$, connecting the center of the camera with the same point M (Figure 4.40).

This technique is widely used in various fields. It is very fast compared to scanning point-by-point or line-by-line. From an experimental point of view, it is robust and not sensitive to ambient light, as we only need to distinguish the white and black patterns, even if the contrast is affected by lighting or by the quality of the object's surface.

4.4.2. Description of the structured light projection measurement setup

The structured light metrology system built at the Photomechanics laboratory has a modular design, allowing for a quick adaptation to surface topography measurements on objects of varying sizes, from several millimeters to one meter. It can also be used as a measurement setup for DIC of in-plane deformation fields, with a resolution suitable for measuring larger scale deformations than the typical range of SI. Two configurations of this setup are shown in Figure 4.41.

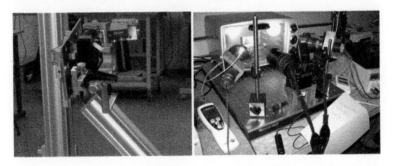

Figure 4.41. *Structured light projection and digital image correlation setups*

4.4.3. Examples of surface topography measurements by structured light projection

The main purpose of these tests, in the context of the AUDACE project, is to estimate the characteristics (sensitivity, noise and errors) of this type of technique when applied to electronic assemblies. Figure 4.42 shows, in a preliminary form, the results that were obtained. The measurement resolution is approximately 1 micron.

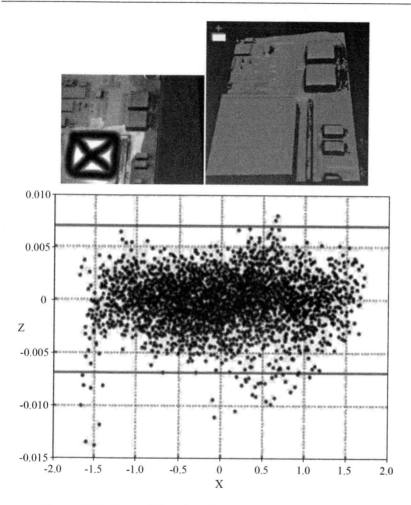

Figure 4.42. *Above: 3D surface topography measurement of a UHF card by structured light projection; below: 3D point cloud (3,600 points) of a 25 mm^2 subregion*

The 3D point cloud allows estimating the measurement uncertainty to approximately 6 microns to a confidence level of 99%.

When the measurement setup is used to measure, through DIC, the in-plane displacements (u, v) produced by thermal stress of the bare controller card, we see that the deformations vary greatly from one region to another, depending on the presence and extent of copper plated areas. This phenomenon hinders quantitative estimation of material properties and the linear expansion coefficient in particular.

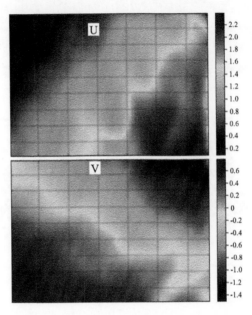

Figure 4.43. *Thermal displacement fields in the plane of the controller card*

4.5. Conclusion

The full field measurement techniques presented in this chapter are well suited for the metrology of thermal or mechanical induced deformations in electronic components and PCBs. Specifically in the case of dynamic metrology, these techniques can provide valuable results for the phenomenological analysis and validation of numerical models used in the study of reliability and physical causes of failure.

4.6. References

[BOR 05] BORZA D.N., "Mechanical vibration measurement by high-resolution time-averaged digital holography", *Measurement Science and Technology*, vol. 16, pp. 1853–1864, 2005.

[BOR 06] BORZA D.N., "Full-field vibration amplitude recovery from high-resolution time-averaged speckle interferograms and digital holograms by regional inverting of the Bessel function", *Optics and Lasers in Engineering*, vol. 44, pp. 747–770, 2006.

[BOR 12a] BORZA D.N., NISTEA I.T., "Experimental investigation by speckle interferometry of solder joint failure under thermomechanical load aggravated by boundary conditions at board level", *Journal of Electronic Packaging*, vol. 134, pp. 041007-1–041007-8, December 2012.

[BOR 12b] BORZA D.N., NISTEA I.T., "High temporal and spatial resolution in time resolved speckle interferometry", *Optics and Lasers in Engineering*, vol. 50, pp. 1075–1083, 2012.

[MOR 08] MOREAU A., BORZA D.N., NISTEA I.T., "Full-field vibration measurement by time-average speckle interferometry and by Doppler vibrometry – a comparison", *Strain*, vol. 44, no. 5, pp. 386–397, 2008.

[NIS 10] NISTEA I.T., Développement des techniques optiques et acoustiques de mesure de champs orientées vers la vibroacoustique, PhD thesis, INSA Rouen, 2010.

[NIS 13] NISTEA I.T., BORZA D.N., "High speed speckle interferometry for experimental analysis of dynamic phenomena", *Optics and Lasers in Engineering*, vol. 51, pp. 453–459, 2013.

5

Characterization of Switching Transistors Under Electrical Overvoltage Stresses

Previous studies of component degradation due to electrostatic discharges (ESD) have led to the installation of protective circuits improving the resistance of the components. These ESD protection circuits have led to other degradation phenomena. The sources of degradation are not easily identified. The phenomena involved are slow, but the energy is sufficient to degrade the component. This chapter presents a method for reproducing these phenomena due to electrical overvoltage stresses. The results obtained by simulation and experimentation of metal–oxide–semiconductor field-effect transistor (MOSFET) switching transistors in an inductive magnetic field environment are analyzed.

5.1. Introduction

Overvoltage stress (OVS), also referred to as electrical OVS (EOVS), appears during the power-up of electronic equipment and during the switching of transistors from the "OFF" state to the "ON" state and vice versa. OVS is distinguished from ESD by the duration of the overvoltage peak and by lower voltages. An ESD lasts approximately 100 ns and can reach several thousand volts. The OVS exceeds the maximum that a component is able to support (absolute maximum rating (AMR)). The durations are longer than 200 ns.

Chapter written by Patrick MARTIN, Ludovic LACHEZE, Alain KAMDEL and Philippe DESCAMPS.

MOSFETs are used to switch power in many electronic applications in the automotive and aircraft industries. Switching OVSs are revealed by the presence of a high-level voltage on one of the pins of the transistor.

This chapter presents a method of characterizing the behavior of MOSFET transistors with regard to OVSs which appear in industrial applications during switching. The principle of the method is to synchronize progressively increased voltage pulses to the drain while sending a command to the gate to change the ON state to the OFF state. The results obtained on three types of silicon MOSFET transistors used in the switching circuits of a direct current (DC)/DC buck converter or in the control circuit of a steering assist Motor are analyzed.

5.2. Stress test over ESD/EOV electric constraints

In order to characterize the behavior of MOSFET transistors when they are subjected to stresses, specific test equipment, a transient power generator (TPG), is set up. The TPG's objective is to produce OVS impulses similar to those found in industrial applications.

5.2.1. *Description of the TPG test equipment*

Figure 5.1 presents a diagram of the architecture of the TPG test equipment. A 25 MHz bandwidth arbitrary function generator is used for generating test signals. The waveforms of the stresses are fully programmable and downloadable. Arbitrary signals with up to 128 K samples and with a maximum sampling frequency of 250 MS/s can be produced. Thus, test signals comply with the characteristics requirements described in the standards.

The amplifiers (Figure 5.2) cover several frequency ranges between continuous current and 10 MHz, voltages of approximately 1,500 V and currents of approximately 8 A. The motherboard connects the power supplies and the stress signals to the pins of the component under test and to the power monitor. It is given specific daughter boards for each component to be tested.

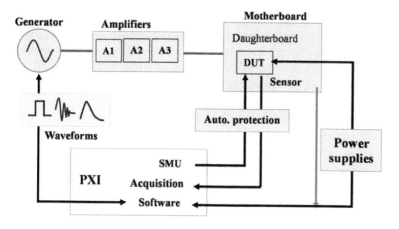

Figure 5.1. *Diagram of the test equipment*

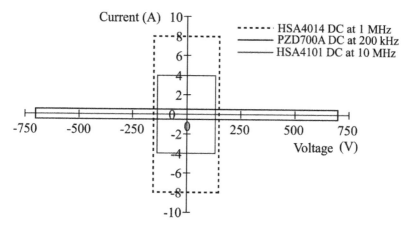

Figure 5.2. *Voltage and current ranges of amplifiers*

Data are acquired by a source meter unit (SMU). This unit uses Peripheral Component Interconnect (PCI) eXtensions for Instrumentation (PXI) technology and is controlled by a software using the Laboratory Virtual Instrument Engineering Workbench (LabView) graphical programming platform. Test signals can be checked visually using an oscilloscope.

5.2.2. *Stresses applied to the transistor*

Three stress waveforms are applied. Each of these waveforms refers to an ESD standard. The number of stress cycles applied to the component is used to study the degradation mechanisms.

The software that is provided by the generator is used to create the waveforms. Signals can also be imported from a digital oscilloscope. The user can also create his/her own signal database by using a spreadsheet or generating files from a couple of lines of code in a programming language.

A square waveform is used to characterize the long-term electric events which may be caused by an incorrect handling of the power supply wires. The degradation mechanisms can be studied by varying the parameters of this waveform (Figure 5.3) [ESD 04].

The damped sine wave described in Figure 5.4 corresponds to the electric stresses caused by an inductive discharge. This wave is defined in the ESD machine model (MM) test standard [JED 12].

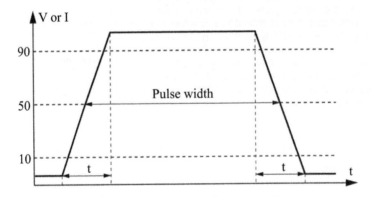

Figure 5.3. *Waveform of the ESD TLP standard*

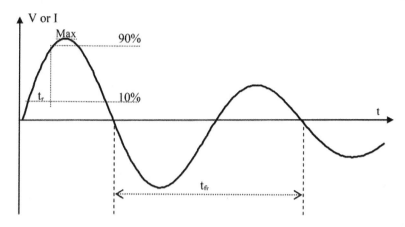

Figure 5.4. *Description of the wave in the ESD MM standard*

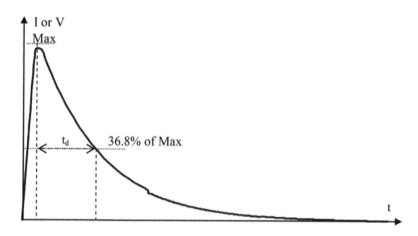

Figure 5.5. *ESD HBM standard wave*

The resistance capacitor (RC) waveform creates electric stresses which are found in a capacitive environment. This waveform is specified by the ISO 7637-2 Pulse2a and ESD HBM test standards [ESD 12] (Figure 5.5). It can be synthesized by a piecewise discrete function. The rise part of the stress signal is linear. The rest of the signal is a decreasing exponential function.

5.2.3. *Testing procedure*

The testing procedure of a component is described in Figure 5.6. The degradation of the component under test is first characterized by measuring the electrical current versus voltage (I(V)) characteristic. An electrical overvoltage stress is then applied and a second I(V) electric characterization is performed. The failure criteria correspond either to a variation of ±1% of the second I(V) characteristic relative to the initial one or to a variation of ±5% of the power consumption of the component. If the observed variations are below these criteria the component is considered undamaged and another stress is applied.

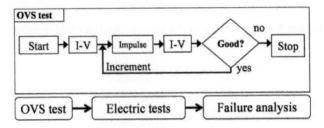

Figure 5.6. *Test flowchart*

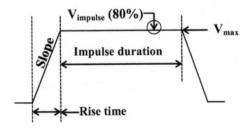

Figure 5.7. *Data acquisition*

The voltage and current measurements are done during the impulse duration. They are performed at 80% of the duration of the impulse, as shown in Figure 5.7.

5.2.4. *TPG capabilities*

These examples show the capabilities of the TPG test equipment to generate or reproduce waveforms which respect the standards (rise times, mid-height width, duration, etc.) or specific criteria.

5.3 Simulation results

5.3.1. *Highlighted phenomena*

When the gate signal of a MOSFET transistor switches (Figure 5.8), an overvoltage appears between the drain and the source (V_{DS}) as well as current transients (I_{DS}).

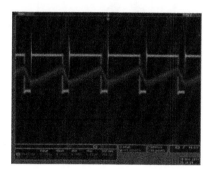

Figure 5.8. *Captures of drain–current transients I_{DS} and drain–voltage transients V_{DS}*

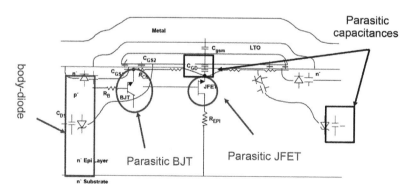

Figure 5.9. *Structure of an MOS transistor and various parasitic effects*

5.3.2. *Influence of parasitic phenomena*

In MOSFETs, there are two types of parasitic phenomena: intrinsic parasitic effects and parasitic effects linked to the environment. Figure 5.9 presents the structure of an MOS and the various parasitic effects. The metallization of the drain and the gate leads to contact resistances. There is also a gate capacitance and a body diode.

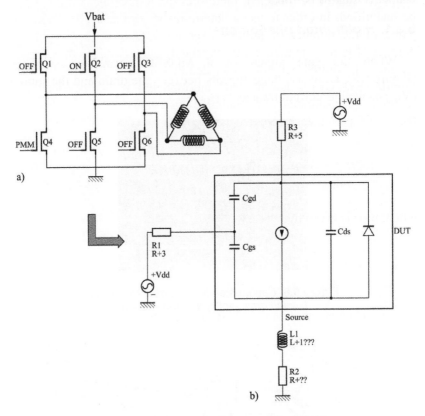

Figure 5.10. *a) Electric diagram of the H bridge and b) equivalent diagram of the tested circuit*

In industrial electric motor control applications, switching transistors are assembled in an H bridge circuit (Figure 5.10(a)). The transistors are controlled by pulsewidth modulations (PWMs). The lengths of the circuits induce an inductance and a resistance. The diagram of the parasitic elements is presented in Figure 5.10(b). V_{GS} voltage is simulated to assess their effect.

Before performing the simulations, the model is validated. The model is satisfactory if the relative error is lower than 5%. The model overestimates the OVSs when the transistor switches from state OFF to state ON. This error is due to the simplified model of the body diode which does not take into account the polarization effects. The values of the resistances and inductances related to the environment also have an impact on V_{GS}. Figure 5.11 represents the evolution of V_{GS} for two values of the drain resistance R_3 (1 and 10 Ω). The mitigation of the peak is in the order of 35% when R_3 increases by a factor of 10.

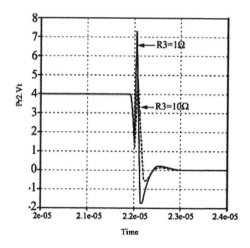

Figure 5.11. *Evolution of the V_{GS} peak as a function of the value of R_3*

Figure 5.12(a) shows that for two values (1 and 10 Ω) of the source resistance R_2, V_{GS} remains constant. However, after the peak, the amplitude of the oscillation is reduced.

In order to characterize the impact of the inductance L_1 on V_{GS} (Figure 5.12(b)), three values of inductance were studied (0.1, 1 and 10 µH). For a variation of L_1 by a factor of 100, the amplitude of the peak increases by 25%.

Figure 5.13 shows that the oscillation of V_{GS} is highly sensitive to the value of the gate resistance R_1 (1 and 10 Ω). It follows that the oscillation of V_{GS} is highly sensitive to the value of R_1.

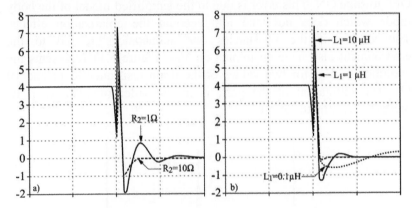

Figure 5.12. a) Evolution of V_{GS} as a function of R_2 and b) evolution of V_{GS} as a function of L_1

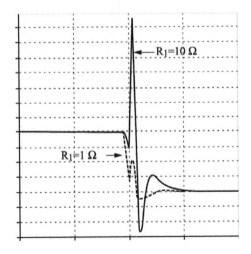

Figure 5.13. Evolution of the V_{GS} peak as a function of the value of R_G

5.4. Experimental setup

The I–V [$I_{DS}(V_{DS})$, $I_{DS}(V_{GS})$, $I_{GS}(V_{GS})$] characteristics determine the evolution of the electric performances of the component when stresses are applied. The parameters are the saturation current (I_{DSAT}), the subthreshold drain–current (I_{DSS}), the threshold voltage (V_{TH}), the resistance of the transistor in its linear region (R_{DSlin}) as well as its transconductance (G_m). The breakdown voltage of the transistor (V_{BR}) is the critical voltage between the drain and the source (V_{DS}) obtained by imposing a current of 250 µA in the drain and by varying the gate voltage.

The applied stress is composed of 10 voltage impulses on the gate of the transistor such that $V_{DS} \geq V_{BR}$. The number of impulses is chosen in order to allow the observation of the degradation of the characteristics $I_{DS}(V_{DS})$. The degradation mechanisms are due to the power dissipated in the component.

If the component is functional, a new stress is applied with a higher voltage V_{DS} followed by a characterization $I_{DS}(V_{DS})$. These steps are repeated until the complete failure of the transistor. After the failure, the circuits are opened either chemically or by laser in order to observe the physical signature of the degradation using an optical microscope.

For the three studied transistors, the stress conditions are summarized in Table 5.1 below:

Transistor	Signal V_{DS}	Step V_{DS}
IR_CR IR-t21: T = 2 µs IR-t22: T = 20 µs IR-t23: T = 200 µs	50% duty cycle T = [2 µs, 20 µs, 200 µs] tr = tf = 100 ns	50 V, 70 V, 90 V, 100 V, 105 V, 110 V, 115 V, 120 V, 125 V
BUK_CX BUK-t2: T = 4 µs BUK-t3: T = 40 µs BUK-t4: T = 400 µs	50% duty cycle T=[4 µs, 40 µs, 400 µs] tr = tf = 100 ns	10 V, 20 V, 30 V, 35 V, 40 V, 42 V

NP110_CE NP110-t2: T = 6 μs NP110-t3: T = 40 μs NP110-t4: T = 400 μs	50% duty cycle T=[6 μs, 40 μs, 400 μs] tr = tf = 100 ns	10 V, 20 V, 30 V, 35 V, 40 V, 42 V

Table 5.1. *Stress conditions*

5.4.1. Measurement results and analysis of observed phenomena

5.4.1.1. V_{BR} measurements of IR_CR transistors

The IR_CR transistors are dedicated to switching electronics. Their AMRs are summarized in Table 5.2.

$I_{DMAX(DC)}$	56 A to 25°C
$I_{DMAX(pulsed)}$	220 A to 25°C
P_{diss}	140 W
V_{GS}	−15 to 15 V
T_{Jmax}	175°C
V_{DSmax}	100 V

Table 5.2. *AMR of IR_CR transistors*

Measuring the V_{BR} by drain–current injection gives the maximum voltage that can be reached by the drain of these three components:

V_{BR} (IR–t21) = 112 V

V_{BR}(IR–t22) = 113 V

V_{BR}(IR–t23) = 113 V

These components are homogeneous over this criterion. Thus, the avalanche effect should not be destructive for $V_{DS} < V_{BR}$.

5.4.1.2. V_{BR} measurements of IR_CR, BUK_CX and NP110_CE transistors

The BUK_CX and NP110_CE transistors are also dedicated to switching electronics. Their AMRs are summarized in Table 5.3.

Parameters	BUK_CX	NP110_CE
$I_{DMAX(DC)}$	100 A to 25°C	110 A to 25°C
$I_{DMAX(pulsed)}$	1,249 A to V_{DS} = 10 V	440 A
P_{diss}	333 W	288 W
V_{GS}	From –20 to 20 V	From –20 to 20 V
T_{Jmax}	175°C	175°C
V_{DSmax}	30 V	30 V

Table 5.3. *AMRs of BUK_CX and NP110_CE transistors*

In contrast to the IR_CR transistor, which has a classical design with a single gate, these two transistors are designed with the TrenchMOS technology (Figure 5.14):

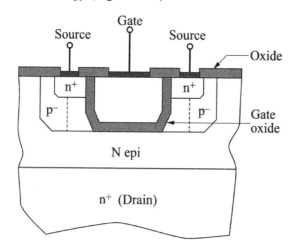

Figure 5.14. *Diagram of the TrenchMOS technology*

The measurements of V_{BR} by drain–current injection are represented in Figures 5.15 and 5.16. The maximum V_{BR} voltage for each of the six components is given in Table 5.4.

V_{BR} (BUK t2) = 39.6 V	V_{BR} (NP110 t2) = 36.9 V
V_{BR} (BUK t3) = 39.6 V	V_{BR} (NP110 t3) = 38 V
V_{BR} (BUK t4) = 39.6 V	V_{BR} (NP110 t4) = 36.8 V

Table 5.4. V_{BR} *voltages measured by drain–current injection*

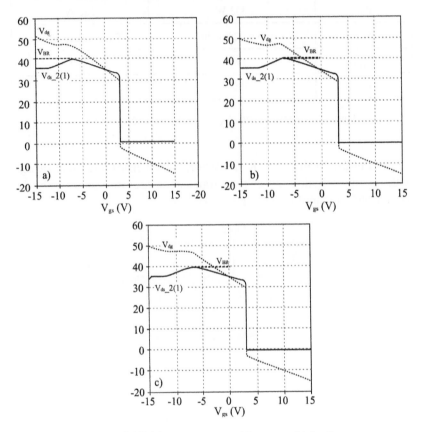

Figure 5.15. V_{BR} *measurement by current injection for the NP110_CE transistor*

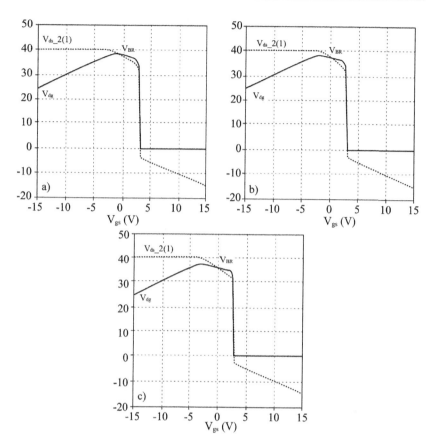

Figure 5.16. V_{BR} measurement by current injection for the BUK_CX transistor

5.4.1.3. Interpretation of the $I_{DS}(V_{DS})$ and $I_{DS}(V_{GS})$ measurements

The evolution of the characteristics $I_{DS}(V_{DS})$ and $I_{DS}(V_{GS})$ is measured after each stress and is compared using the following parameters:

– saturation current, I_{DSAT};
– subthreshold current of the transistor, I_{DSS};
– threshold voltage, V_{TH};

– transconductance, G_m;

– resistance in linear mode, $R_{ON|LIN}$.

The characteristic $I_{DS}(V_{DS})$ of the components IR-t21, IR-t22, BUK-t2, BUK-t3, BUK-t4 as well as NP110-t2, NP110-t3 and NP110-t4 are studied. An increase in I_{DSAT} as well as a decrease in $R_{ON|LIN}$ is observed in Figures 5.17, 5.18, 5.21 and 5.22. However, when V_{DS} is lower than V_{BR}, these parameters are stable. V_{BR} is an indicator of degradation. The voltage V_{BR} measured by drain–current injection represents the avalanche voltage of the structure diode.

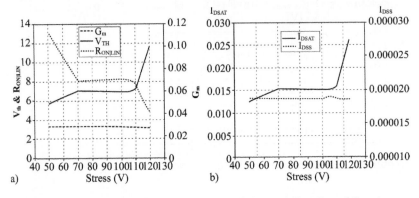

Figure 5.17. Evolution of various characteristics as a function of the stress voltage for the IR-t21 transistor: a) $R_{ON|LIN}$, G_m and V_{TH} and b) I_{DSAT}, I_{DSS}

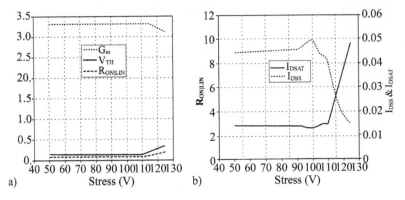

Figure 5.18. Evolution of various characteristics as a function of the stress voltage for the IR-t22 transistor: a) $R_{ON|LIN}$, G_m and V_{TH} and b) I_{DSAT}, I_{DSS}

The threshold voltage V_{TH}, the sub-threshold current IDSS and the transconductance (G_m) are obtained from the characteristic $I_{DS}(V_{GS})$. The current I_{DSS} (at $V_{DS} = 0$ V) does not depend on the stress applied.

From the evolution of these parameters, we can localize the physical degradation.

The drain–current equation of MOS is:

$$I_{DSAT} = W_g \cdot (V_{GS} - V_{TH}) \cdot C_{ox} \cdot v_S \qquad [5.1]$$

where W_g is the width of the gate, C_{ox} is the capacitance of the gate and v_s is the speed of the carrier.

Figure 5.18 shows a decrease in V_{TH}, leading to an increase in the V_{GS}–V_{TH} term and therefore an increase in I_{DSAT}.

Equation [5.2] expresses the threshold voltage that needs to be applied on the gate in order to neutralize the charges.

$$V_{TH} = V_{FB} + 2\varphi_B + \frac{\sqrt{4\varepsilon_s q N_A}}{C_{OX}} \qquad [5.2]$$

where V_{FB} is the voltage of the flat strip, Φ_B is the height of the potential barrier of the metal–oxide–Si contact of the gate and ε_S is the permittivity of silicon. N_A is the dopant density of the p-zone of the channel. A decrease in N_A leads to a decrease in V_{TH}. When $V_{DS} = V_{BR}$, the p-zone is completely depleted, which leads to a zero charge within this zone.

Under the influence of a strong electric field between the drain and the source, the voltage of the body diode is blocked since its depletion region charge can no longer be extended. An impact ionization mechanism is triggered, degrading the doping in the p-zone. The doping becomes weaker than before the application of the stress. This provokes a decrease in V_{TH} and therefore an increase in the I_{DS}.

The triggering of the avalanche phenomenon is confirmed by the chronograms (Figures 5.19, 5.20, 5.23 and 5.24). When V_{DSS} is higher than or equal to V_{BR}, V_{DS} decreases significantly.

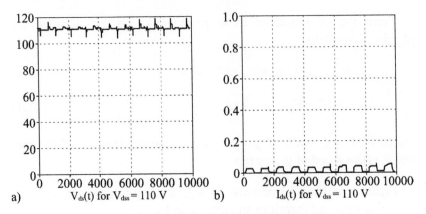

Figure 5.19. *Chronogram of $V_{DS}(t)$ and $I_{DS}(t)$ before the avalanche regime is triggered*

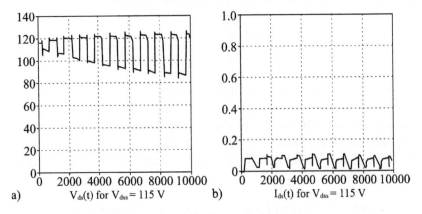

Figure 5.20. *Chronogram of V_{DS} and I_{DS} after the avalanche regime is triggered*

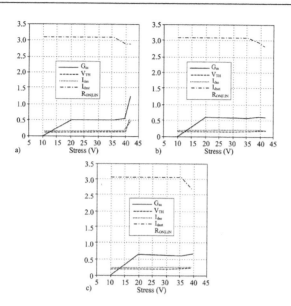

Figure 5.21. Evolution of various characteristics ($R_{ON|LIN}$, G_m, V_{TH}, I_{DSAT} and I_{DSS}) as a function of the stress voltage for the transistor: a) TRANS1-t2, b) TRANS1-t3 and c) TRANS1-t4

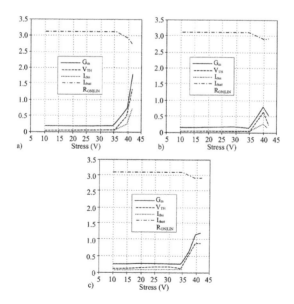

Figure 5.22. Chronogram of $V_{DS}(t)$ and $I_{DS}(t)$ before the avalanche regime is triggered

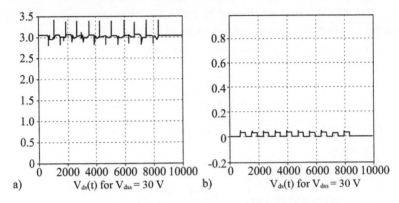

Figure 5.23. Chronogram of $V_{DS}(t)$ and $I_{DS}(t)$ after the avalanche regime is triggered

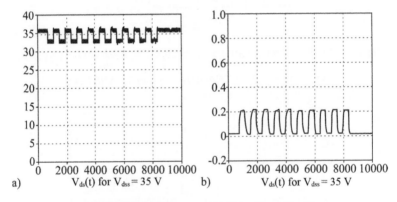

Figure 5.24. Chronogram of $V_{DS}(t)$ and $I_{DS}(t)$ after the avalanche regime is triggered

5.5. Conclusion

This study reproduces the triggering of the avalanche regime in power transistors used in switching electronics driving inductive loads. This mechanism is produced by an overvoltage on the drain caused when the transistor switches from the "ON" state to the "OFF" state. This provokes a brutal variation of the current in the inductance and induces an overvoltage which leads to the complete depletion of the structure diode, causing the avalanche regime.

This avalanche mechanism can lead to the destruction of the transistor by thermal runaway if the phenomenon has sufficient time to reach a high enough temperature. This study is carried out on a transistor having a single gate and two transistors having a gate matrix. In both case studies, the mechanism is triggered in the same manner and leads to a decrease of dopant in the p-zone of the body diode and in the channel.

5.6. References

[ESD 04] ESDA Standard Practice, Electrostatic Discharge (ESD) sensitivity testing Transmission Line Pulse (TLP) – Component level, ANSI SP5.5.1, 2004.

[ESD 12] ESDA–JEDEC Standard, Electrostatic discharge (ESD) sensitivity testing Human Body Model (HBM) – Component level, ANSI JS-001-2012, 2012.

[JED 12] JEDEC Standard, Electrostatic discharge (ESD) sensitivity testing Machine Model (MM), EIA/JESD22-A115, 2012.

6

Reliability of Radio Frequency Power Transistors to Electromagnetic and Thermal Stress

This chapter is focused on the study of the reliability of radio frequency (RF) power transistors used in high power amplifier (HPA) boards. These transistors constitute the central element of (Tx) transmission modules in radar applications. The behavior of RF transistors using gallium nitride (GaN) technology is tested by applying stresses caused by radiated electromagnetic waves, RF signals or thermal exposure. These stress tests lead to failure mechanisms representative of degradations observed in operation. The principle of this stress test approach is to aggravate these degradations by applying stresses that are representative of the harsh environments of RF power transistors.

6.1. Introduction

Microwave power modules are often located in a very harsh electromagnetic environment which disrupts their operation (high temperatures and voltages, and significant RF powers). This environment leads to stresses which degrade the performance of the modules. In this chapter, we study the effect of environmental stresses on the behavior of the characteristics of high electron mobility transistor (HEMT) transistors using GaN technology.

Chapter written by Samh KHEMIRI and Moncef KADI.

In this chapter, the transistor under test is presented, as well as the assembly of the RF power amplifier which is used as a "test support" to characterize and apply stresses. The impact of the electromagnetic, RF and thermal stresses on the characteristics of the RF transistor is then studied. Hypotheses are proposed as to the causes of failure. Failure mechanisms specific to these new technologies are highlighted. Finally, the effects of combined stresses on the behavior of the HEMT transistor are presented.

6.2. The GaN technology

The component studied here (Figure 6.1) is an AlGaN/GaN HEMT transistor. It has a maximum output power of 50 W, a maximum operating frequency of 4 GHz, a maximum drain bias voltage of 28 V, a gate voltage between –3 and 3 V and a gain of around 12 dB. Its substrate is a high-resistance silicon floating layer 150 μm thick. An undoped GaN layer is deposited on this silicon layer. A transition layer adapts the differential thermal expansion and the lattice mismatch between the array of the GaN and the Si. A GaN buffer layer is deposited on the transition layer. The transistor is characterized by a gate perimeter of 16 μm, a gate length of 0.5 μm, a gate-source spacing of 1 μm and a gate-drain spacing of 3 μm [NIT 08].

Figure 6.1. *Inside view of the HEMT NBTP00050 transistor [NIT 10]*

This transistor is included in a microwave power amplifier circuit (Figure 6.2). This circuit is used for testing. It makes it possible to measure the static characteristics (I_{ds}, g_m, etc.) as well as the RF characteristics (gain and drain efficiency). The amplifier is composed of two subcircuits. The first subcircuit polarizes the transistor

(Vds and Vgs). The second subcircuit adapts the RF channels. This amplifier has the following characteristics:

- a central frequency of 3 GHz;
- a maximum output power of 43 W;
- a drain bias voltage of 28 V;
- a drain current of 2.3 A;
- a gain of 11.12 dB;
- a 60% power efficiency.

The transistor is polarized by a voltage Vds = 28 V and a voltage Vgs = –1.35 V, in other words with a quiescent current Ids = 300 mA. These conditions correspond to an AB class amplification since the gate-source voltage is slightly higher than the threshold voltage.

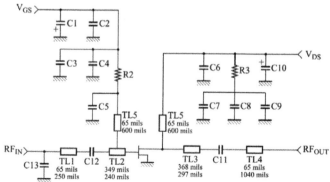

Figure 6.2. *Test circuit: microwave power amplifier*

6.3. Radiated electromagnetic stress

This reliability study of the AlGaN/GaN HEMT is based on electromagnetic stress in radiated mode. The principle is to inject a disruptive power using a magnetic probe (magnetic loop) located at a height "h" above the component being tested [KHE 11b].

6.3.1. *Presentation of the test equipment*

The test equipment (Figure 6.3) is composed of four subsystems.

The first subsystem, which supplies power to the transistor (pulsed or continuous mode), includes the following devices [KHE 11c]:

– an RF signal generator which generates the input signal and has a frequency range between 1 and 20 GHz and an output power between –20 and 15 dBm;

– a power amplifier which amplifies the RF signal (gain of 40 dB and a frequency range between 0.8 and 4.2 GHz).

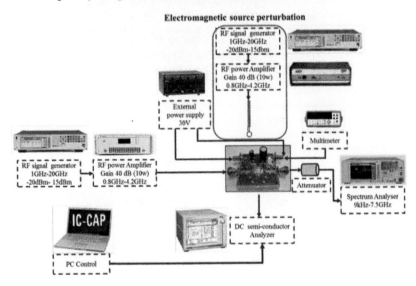

Figure 6.3. *Electromagnetic test equipment. For a color version of this figure, see www.iste.co.uk/elhami/embedded1b.zip*

The second subsystem is composed of all the devices used to polarize the transistor. It has two continuous power supplies in order to deliver the voltage to the drain and the gate.

The third subsystem includes the devices to measure the evolution of the characteristics of the transistor during and after the application of the stress. These computer controlled devices are composed of:

– a multimeter to measure the drain current I_{ds};

– a spectrum analyzer to measure the output power of the transistor;

– a direct current (DC) analyzer (Agilent B1500A) controlled by the IC-CAP software program to measure all the static characteristics of the component under test.

The fourth subsystem is composed of the equipment that generates the disruptive electromagnetic signal:

– a signal generator with an output power between –20 and 15 dBm, and a frequency between 1 and 20 GHz;

– a power amplifier with a frequency between 0.8 and 4.2 GHz, and an amplification gain of 40 dB;

– an electromagnetic probe located above the transistor. The probe chosen for our tests is a magnetic loop with a diameter of 4 mm.

6.3.2. *Results and analysis*

The HEMT transistor is aged in the following conditions:

– the input power of the amplifier is a pulsed signal with a period of 500 μs and a duty cycle of 50%. The input power amplitude (P_{input}) at the frequency f = 3 GHz is: P_{input} = 30 dBm;

– the probe is placed at a height of 2 mm above the gate (to ensure a correct coupling between the probe and the printed circuit board (PCB) tracks) (Figure 6.4);

– P_{probe} = 40 dBm at the frequency f = 3 GHz;

- V_{ds} = 28 V;
- I_{ds} = 760 mA;
- the duration of the tests is 4 hours.

Figure 6.4. *Positioning the probe above the transistor*

	Before stress	After stress	Difference (%)
I_{dmax} (mA)	369.7	311.15	16
R_{ds} (Ohm) (V_{ds} = 28 V)	76.17	90.41	18
V_{th} (V)	−1.55	−1.5	3
$g_{m\,max}$(S)	2.19	1.94	11
Gain (dB)	11.7	9.7	13
D.E(%)(P_{in} = 30dBm)	32	22	30

Table 6.1. *Comparison of the static characteristics before and after application of the electromagnetic stress*

Table 6.1 presents a comparison of the static characteristics before and after application of the electromagnetic stress on an AlGaN/GaN HEMT transistor. Figure 6.5 shows a decrease in the drain current of about 15%. The threshold voltage does not evolve in a significant manner. The transconductance decreases by 11% with a simultaneous decrease in the curves.

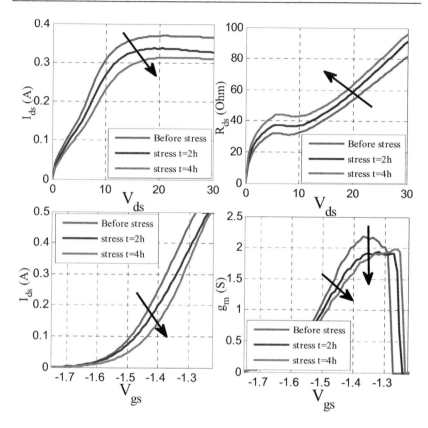

Figure 6.5. *Evolution of the static characteristics before and after application of the electromagnetic stress. For a color version of this figure, see www.iste.co.uk/elhami/embedded1b.zip*

The same results can be seen in [CAI 07] and the same degradation of HEMT characteristics is observed.

Figure 6.6 shows the RF characteristics relative to the input power, before and after the application of the electromagnetic stress. It can be observed that the gain in power drops significantly (2 dB). The efficiency of the drain degrades for input powers higher than 10 dBm. The failure criterion is defined by a reduction of the output characteristic by 10% with respect to its initial value [BRO 04, CRA 06, ALA 11].

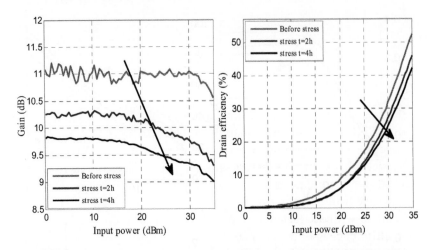

Figure 6.6. *Evolution of the gain and the efficiency of the drain as a function of the input power before and after application of the electromagnetic stress. For a color version of this figure, see www.iste.co.uk/elhami/embedded1b.zip*

Further tests show that these degradations are negligible when the power is lower than 18 dBm. They are also negligible when the probe is placed above the housing, above the drain or at a height greater than 5 mm.

A possible cause of the decrease in the drain current is the width of the current channel, which influences the decrease in the density of the two-dimensional electron gas (2DEG). This is linked to the electron-trapping phenomenon in the buffer layer and in the surface of the AlGaN, and with the increasing resistance of the channel between the gate and the drain. During the power injection through the magnetic loop, an electromagnetic coupling between this loop and the input of the gate is formed. Through this coupling, the power of the electromagnetic perturbation is added on the input power. The carriers are stimulated and fill the existing traps in the structure of the HEMT. The electrons acquire energy and form a second gate in the region between the gate and the drain. This is associated with the electron trapping in the buffer layer, at the interface of the depletion zone of the 2DEG channel and with the increase in the resistance of the channel between the gate and the drain [JOH 11].

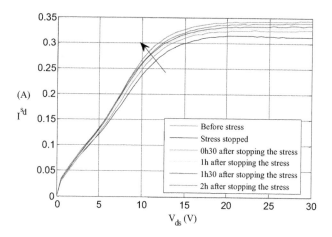

Figure 6.7. *Evolution of the drain current after stopping the stress. For a color version of this figure, see www.iste.co.uk/elhami/embedded1b.zip*

Figure 6.7 shows that after stopping the application of the stress, all the characteristics of the transistor tend to return to their initial states. This is a relaxation phenomenon. The electrons trapped in the buffer zone are freed and return to their initial positions. The return time to equilibrium is 105 min. This observation supports the hypothesis that the electrons responsible for the degradations are trapped in the buffer layer of the AlGaN.

6.4. RF CW continuous stress

The RF stress consists of injecting continuous power waves (CWs) to the transistor input over a four hour period. The amplifier functions in a continuous wave state. Set frequency is 3 GHz and output power is 30 dBm (decibels relative to a milliwatt).

6.4.1. Presentation of the test equipment

The equipment for testing the impact of continuous stresses consists of the above test equipment from which the part that generates the radiated perturbation has been removed (Figure 6.8).

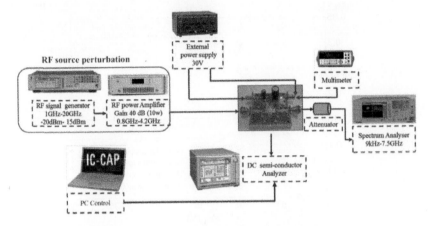

Figure 6.8. *RF test equipment. For a color version of this figure, see www.iste.co.uk/elhami/embedded1b.zip*

	Before stress	After stress	Difference (%)
I_{dmax} (mA)	369.7	302	18
R_{ds} (Ohm) (V_{ds} = 28 V)	76.17	83.7	9
V_{th} (V)	−1.55	−1.49	2
$g_{m\ max}$(S)	2.19	1.83	16
Gain (dB)	11.7	10.25	8
D.E(%)(P_{in} = 30 dBm)	32	29	7

Table 6.2. *Comparison of the static characteristics before and after the RF stress*

6.4.2. *Results and analysis*

The transistor is subjected to the following test conditions:

– P_{input} = 30 dBm (CW) at a frequency of f = 3 GHz;

– V_{ds} = 28 V;

– Ids = 1.2 A;

– test duration: 4 hours.

Table 6.2 presents the static characteristics before and after the application of the stress. The drain current drops by 18% with respect to its initial value. The other parameters do not have significant variations. The main cause of this degradation is the electron-trapping phenomenon in the AlGaN buffer layer [JOH 10, JOH 11].

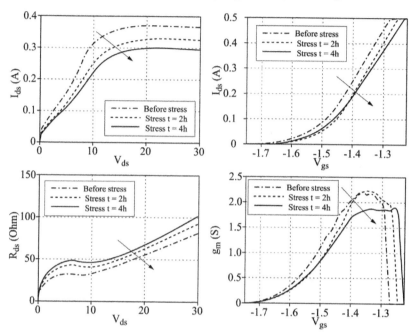

Figure 6.9. *Evolution of the static characteristics before and after RF stress*

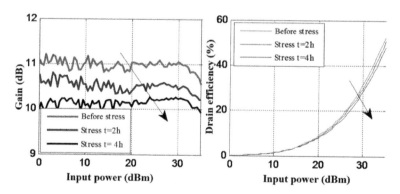

Figure 6.10. *Evolution of the gain and the drain efficiency relative to the input power before and after application of the RF stress. For a color version of this figure, see www.iste.co.uk/elhami/embedded1b.zip*

6.5. Thermal exposure

6.5.1. *Presentation of the test equipment*

The thermal exposure test equipment (Figure 6.11) includes the RF test equipment in which an oven with an adjustable temperature between –150 and 250°C is added.

The principle of this thermal test is to inject a pulsed power at the input of the microwave power amplifier at various temperatures. The conditions are the following:

– temperature: 90 or –40°C;

– P_{input} = 30 dBm, pulsed, at a frequency of 3 GHz. This power has a period of 500 μs and a duty cycle of 50%;

– V_{ds} = 28 V;

– I_{ds} = 760 mA;

– Test duration: 8 h.

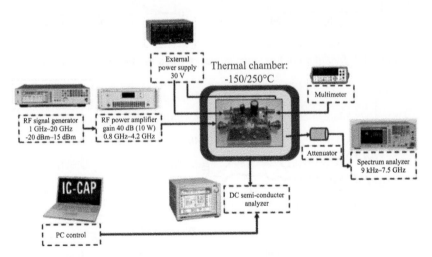

Figure 6.11. *Thermal exposure testbed. For a color version of this figure, see www.iste.co.uk/elhami/embedded1b.zip*

6.5.2. Results and analysis

6.5.2.1. Study at temperature T = 90°C

Figure 6.12 shows the characteristics of the HEMT after 8 h of application of the thermal stress at a temperature of 90°C. We observe a degradation of the drain current. Its amplitude shows a reduction of 114 mA. The resistance of the drain increases by 32 Ω. The transconductance decreases by 0.4 S.

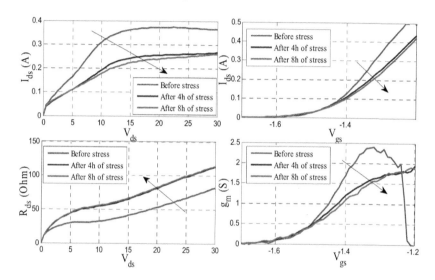

Figure 6.12. *Evolution of the static characteristics before and after application of the thermal stress, 90°C. For a color version of this figure, see www.iste.co.uk/elhami/embedded1b.zip*

The threshold voltage shift is explained by the activation of deep centers situated at the interfaces of the AlGaN structure. A deep trench in a semiconductor behaves as a trap, in other words as a recombination center or a generation center. The trench is considered to be a trapping site if there is re-emission of the trapped carrier toward the origin strip after a characteristic trapping time. In the case where a carrier of opposite sign is trapped before the first one is re-emitted, the site acts as a recombination center. If the deep centers are distributed in a homogeneous manner in the channel volume, their charge state does not influence the global charge density. No matter the state of the deep centers (empty or full, neutral or ionized), the threshold voltage remains unchanged. If these traps are situated near an interface, they may create, as a function of their charge states, a parasite depletion in the channel. If the traps are situated at the buffer/channel interface, at low temperature, they also create a parasitic depletion at the rear of the channel. This means that the threshold voltage is lower if the traps are empty (parasitic depletion removed). The degradation of the current and the transconductance at high temperature is due to the degradation of the mobility of the 2DEG channel. At high temperature, the increase in the internal temperature of the components causes the decrease in the thermal conductivity of the materials. This reduction affects the 2DEG density.

6.5.2.2. *Study at temperature T = –40°C*

After 8 hours of applying the stress at –40°C, we observe the opposite phenomenon than the phenomenon observed at a temperature of 90°C. The drain current increases by 70 mA. The resistance of the drain decreases by 17 Ω. The transconductance increases by 1 S (Figure 6.14). The gain of the RF amplifier increases by 10% (Figure 6.15). The efficiency of the drain does not change.

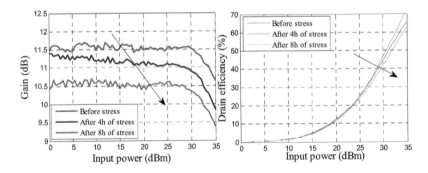

Figure 6.13. *Evolution of the gain and the drain efficiency as a function of the input power before and after application of the thermal stress, 90°C. For a color version of this figure, see www.iste.co.uk/elhami/embedded1b.zip*

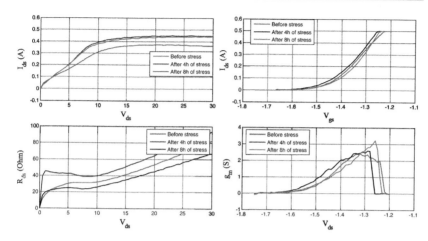

Figure 6.14. *Evolution of the static characteristics before and after application of the thermal stress, T = –40°C. For a color version of this figure, see www.iste.co.uk/elhami/embedded1b.zip*

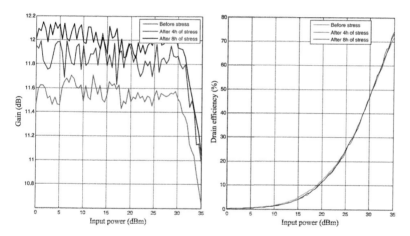

Figure 6.15. *Evolution of the gain and the drain efficiency after application of the thermal stress, T = –40°C. For a color version of this figure, see www.iste.co.uk/elhami/embedded1b.zip*

6.6. Combined stresses: RF CW + electromagnetic (EM) and electric + EM

In this section, we will look at the combination of the electromagnetic stress with the other stresses (electric and RF).

6.6.1. *Effect of the simultaneous application of EM and RF stresses*

The electromagnetic stress is combined with the RF stress. Disruptive power is applied to the input of the probe. At the same time continuous power is applied to the transistor input. The test conditions are:

– P_{input} = 30 dBm, this is a continuous power, at frequency f = 3 GHz;

– P_{probe} = 40 dBm at frequency f = 3 GHz;

– V_{ds} = 28 V;

– I_{ds} = 1.2 A;

– test duration: 4 hours.

Figures 6.16 and 6.17 present the variation of the static and RF characteristics after 4 hours of application of the combined "EM + RF" stresses.

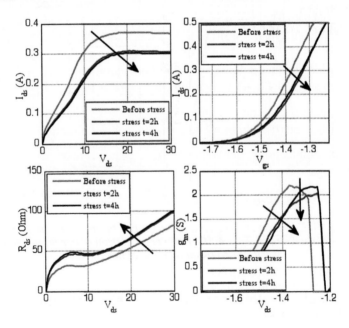

Figure 6.16. *Evolution of the static characteristics before and after application of the "RF + EM" stress. For a color version of this figure, see www.iste.co.uk/elhami/embedded1b.zip*

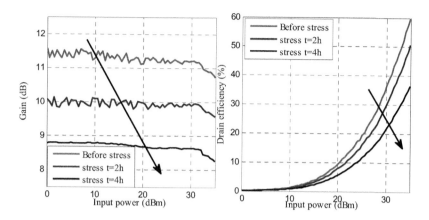

Figure 6.17. *Evolution of the gain and drain efficiency as a function of the input power before and after application of the "RF+ EM" stress. For a color version of this figure, see www.iste.co.uk/elhami/embedded1b.zip*

	Before	RF stress	EM stress	EM + RF stress
I_{dmax} (mA)	369.7	304.8	311.15	296.7
R_{ds} (Ω) ($V_{ds} = 28$ V)	76.13	91.83	90.41	93.49
$g_{m\,max}$(S)	2.39	1.93	1.94	1.84
V_{th} (V)	−1.55	−1.51	−1.5	−1.46
Gain (dB)	11.5	9.4	9.7	8.7
D.E(%)($P_{in} = 30$dBm)	36.37	25.25	22	21.81

Table 6.3. *Comparison of the static characteristics before and after the combined stress: RF + EM*

Table 6.3 compares the static characteristics before and after application of the simultaneous RF and EM stresses. The drain current amplitude shows a reduction of 73 mA. The drain current degrades by 64 mA when we apply an RF stress and by 58.55 mA when we apply an electromagnetic stress. An offset of the threshold voltage of around 0.09 V is observed. The gain in power of the amplifier degrades by 2.8 dB in the case of simultaneous stresses, by 1.4 dB in the case of an RF stress and by 1.8 dB in the case of the electromagnetic stress. The application of simultaneous stresses has little effect on the drain efficiency.

6.6.2. Effect of the simultaneous application of electromagnetic and continuous DC stresses

This test consists of applying the electromagnetic stress while polarizing the HEMT by a voltage Vgs equal to 0 V instead of −1.35 V. The conditions are the following:

– P_{input} = 30 dBm, pulsed, at a frequency f = 3 GHz. The period of this power is 500 µs and has a duty cycle of 50%;

– P_{probe} = 40 dBm at the frequency f = 3 GHz;

– V_{ds} = 28 V;

– V_{gs} = 0 V;

– I_{ds} = 2.1 A;

– Test duration: 4 hours.

Figure 6.18 shows the evolution of the output characteristics before and after application of the DC+ EM stresses. Figure 6.19 shows the evolution of the transfer characteristics. Figure 6.20 shows the power gain and the drain efficiency as a function of the input power before and after four hour application of combined DC and EM stresses.

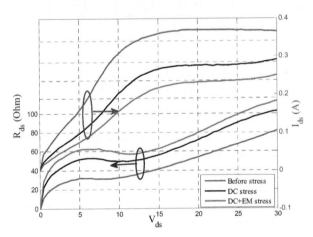

Figure 6.18. *Evolution of the output characteristics before and after application of the DC+ EM stresses. For a color version of this figure, see www.iste.co.uk/elhami/embedded1b.zip*

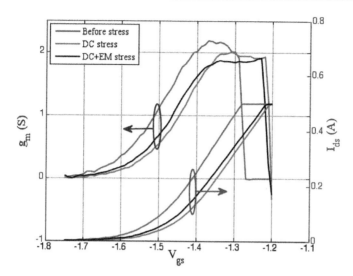

Figure 6.19. *Evolution of the transfer characteristics before and after application of the DC+ EM stresses. For a color version of this figure, see www.iste.co.uk/elhami/embedded1b.zip*

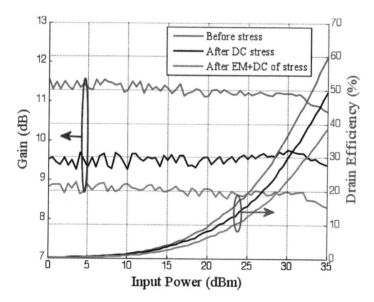

Figure 6.20. *Evolution of the RF characteristics before and after application of the DC+ EM stresses. For a color version of this figure, see www.iste.co.uk/elhami/embedded1b.zip*

	Before stress	DC stress	EM stress	EM+DC stress
I_{dmax} (mA)	369.7	293.8	311.15	242.1
R_{ds} (Ω) (V_{ds} = 28 V)	76.13	98.73	90.41	108.3
$g_{m\,max}$ (S)	2.39	2.03	1.94	1.84
V_{th} (V)	−1.55	−1.486	−1.5	−1.459
Gain (dB)	11.5	9.45	9.7	8.7
D.E(%)(P_{in} = 30 dBm)	36.37	29.81	22	23.27

Table 6.4. *Summary of the static characteristics before and after application of the DC+EM stresses*

Table 6.4 gives the static characteristics before and after application of the DC+EM stress. The application of 4 hours of DC+EM stresses causes significant degradations of the characteristics of the HEMT transistor. The maximum currents of the drain and gain are reduced, respectively, by 128 mA and 2.8 dB. The maximum current of the drain shows a reduction of 75 mA after application of the electric stress and 59 mA after application of the electromagnetic stress. The gain degrades by 2 dB.

6.7. Conclusion

In this chapter, the effect of environmental stresses on the characteristics of HEMT transistors using GaN technology is studied. A piece of test equipment is developed in order to apply electromagnetic, electric, RF and thermal stresses either individually or combined. This test equipment operates at a frequency up to 4 GHz. It has an output power of 50 W.

A 10% degradation of the characteristics of the HEMT transistors is observed, which signifies that the component has failed.

The effect of combined stresses is more significant. The electromagnetic and electric stresses degrade the drain current. The observed degradations are caused by trapping phenomena in the structure of the transistor.

6.8. References

[AKT 96] AKTAS O., FAN Z.F., MOHAMMAD S.N. et al., "High temperature characteristics of AlGaN/GaN modulation doped field effect transistors," *Applied Physics Letters*, vol. 69, no. 25, pp. 3872–3874, December 1996.

[ALA 09] ALAEDDINE A., KADI M., DAOUD K. et al., "Effects of electromagnetic near-field contrainte on SiGe HBT's reliability", *Microelectronics Reliability Journal*, vol. 49, nos 9–11, pp. 1029–1032, 2009.

[ALA 11] ALAEDDINE A., Le Transistor Bipolaire à Hétérojonction Si/SiGe sous contraintes électromagnétiques: des dégradations électriques à l'analyse structurale, PhD thesis, University of Rouen, February 2011.

[ARU 02] ARULKUMARAN S., EGAWA T., ISHIKAWA H. et al., "High-temperature effects of AlGaN/GaN high-electron-mobility transistors on sapphire and semi-insulating SiC substrates", *Applied Physics Letters*, vol. 80, no. 12, pp. 2186–2188, March 2002.

[ATR 09] ATROUS S., Mise en place d'une méthodologie de caractérisation en immunité champ proche de dispositifs électroniques, PhD thesis, University of Rouen, January 2009.

[BAU 07] BAUDRY D., ARCAMBAL C., LOUIS A. et al., "Applications of the near-field techniques in EMC investigations", *IEEE Transactions on Electromagnetic Capability*, vol. 49, pp. 485–493, 2007.

[BOR 01] BORGARINO M., MENOZZI R., CATTANI D. et al., "Reliability physics of compound semiconductor transistors for microwave", *Microelectronic Reliability Journal*, vol. 41, 2001.

[BOU 09] BOULINGUI S.A., Etude du couplage électromagnétique entre circuits intégrés par émulation du perturbateur– Application en téléphonie 3G, PhD thesis, Paul Sabatier University, November 2009.

[BRO 04] BROWN J.D., NAGY S., SINGHAL S. et al., "Performance of AlGaN/GaN HFETs fabricated on 100 mm silicon substrates for wireless base station applications", *IEEE MTT-S International Microwave Symposium Digest*, pp. 833–836, 2004.

[CAI 07] CAI Y., CHENG Z., YANG Z. et al., "High-temperature operation of AlGaN/GaN HEMTs direct-coupled FET logic (DCFL) integrated circuits", *IEEE Electron Device Letters*, vol. 28, no. 05, pp. 328–331, May 2007.

[CRA 06] CRAIG G., ARNOLD T., MARTIN L. et al., "Evaluation of SiGe: C HBT intrinsic reliability using conventional and step constraint methodologies", *Microelectronics and Reliability*, vol. 46, pp. 1272–1278, 2006.

[DE 06] DEMARTY S., Contribution à l'étude électromagnétique théorique et expérimentale des cartes de circuit imprimé, PhD thesis, University of Limoges, August 2006.

[HER 05] HER J.C., LEE K.M., LEE S.C. et al., "Pulsed current-voltage-temperature characteristics of AlGaN/GaN high electron mobility transistor under isothermal conditions", *Japanese Journal of Applied Physics*, vol. 44, no. 4B, pp. 2726–2728, 2005.

[JOH 07] JOH J., XIA L., DEL ALAMO J.A., "Gate current degradation mechanisms of GaN high electron mobility transistors", *IEEE International Meeting of Electron Devices 2007(IEDM 2007)*, 10–12 December 2007.

[JOH 10] JOH J., ALAMO J.A., LANGWORTHY K. et al., "Role of contrainte voltage on structural degradation of GaN high electron mobility transistors", *Microelectronic Reliability Journal*, August 2010.

[JOH 11] JOH J., ALAMO J.A., "A current-transient methodology for trap analysis for GaN high electron mobility transistors", *IEEE Transaction on Electron Devices*, vol. 58, no. 1, January 2011.

[KHE 11a] KHEMIRI S., KADI M., LOUIS A. et al., "Study of electromagnetic field contrainte impact on AlGaN/GaN HEMT transistor performances", *German Microwave Conference (GeMiC 2011)*, Darmstadt, Germany, 2011.

[KHE 11b] KHEMIRI S., LOUIS A., MAZARI B., "Effects of electromagnetic near-field contrainte on DC and RF performances of AlGaN/GaN HEMT", *8th International Conference on Computation in Electromagnetics (CEM 2011)*, Wroclaw, Poland, 2011.

[KHE 11c] KHEMIRI S., KADI M., LOUIS A., "Evolution des performances d'un amplificateur de puissance hyperfréquence à base de transistor HEMT AlGaN\GaN sous des contraintes RF et électromagnétique", *Journées Nationales Micro-onde*, Brest, France, 2011.

[KHE 11d] KHEMIRI S., KADI M., LOUIS A., "Reliability study of AlGaN/GaN HEMT under electromagnetic contrainte", *22nd European Symposium on Reliability of Electron Devices, Failure Physics and Analysis*, Bordeaux, France, 2011.

[KIK 04] KIKKAWA T. et al., "An over 200-W output power GaN HEMT pushpull amplifier with high reliability", *IEEE International Microwave Symposium*, pp. 1347–1350, 2004.

[LAM 06] LAMOUREUX E., Étude de la susceptibilité des circuits intégrés numérique aux agressions hyper-fréquences, PhD thesis, Institut national des sciences appliquées de Toulouse, January 2006.

[MAE 01] MAEDA N., TSUBAKI K., SAITOH T. et al., "High-temperature electron transport properties in AlGaN/GaNheterostructures," *Applied Physics Letters*, vol. 79, no. 11, pp. 1634–1636, September 2001.

[MAT 07] MATSUSHIT A., TERAMOTO S., "Reliability study of AlGaN/GaN HEMTs device", *CS MANTECH Conference*, Austin, TX, 2007.

[NIT 08] NITRONEX CORPORATION, GaN essentials: Substrates for GaN RF devices, Application Notes 011, June 2008.

[NIT 10] NITRONEX CORPORATION, NPTB00050 Qualification Document: NGD010, 2010.

[OKA 04] OKAMOTO K. et al., "Improved power performance for a recessed-gate AlGaN/GaN heterojunction FET iith a field-modulating plate", *IEEE Transactions on Microwave Theory and Techniques*, vol. 52, p. 2536, 2004.

[PAR 09] PARK S.Y., FLORESCA C., JIMENEZ J.L. et al., "Physical degradation of GaN HEMT devices under high drain bias reliability testing", *Microelectronics Reliability*, 2009.

[RAM 01] RAMAKRISHNA V., NAIQAIN Q.Z., STACIA K., "The impact of surface states on the DC and RF characteristics of AlGaN/GaN HFETs", *IEEE Transaction Electron Devices*, pp. 560–566, 2001.

[TRE 91] TREW R.J., BIKBRO L., "Gate breakdown in MESFETs and HEMT's", *IEEE Electron Device Letters*, vol. 12, pp. 524–526, October 1991.

[TRE 05] TREW R.J., BILBRO G.L., "Large-signal IMPATT-mode operation of AlGaN/GaN HFET's", *German Microwave Conference(GeMiC 2005)*, 2005.

7

Internal Temperature Measurement of Electronic Components

This chapter presents the results of a study in which the temperature and the micro-displacements of the chip surface of high-frequency power electronic components used in radar and telecommunication systems are measured. Several techniques are applied. Their advantages, disadvantages and shared fields of application are discussed. Results from several samples show that the different approaches converge. The originality of this study is that the measurements of chip surface temperature and displacement are obtained simultaneously. This approach makes it possible to then calculate the thermal resistance of an electronic component and characterize the evolution of this resistance over component lifetime.

7.1. Introduction

The fields of radar and radio frequency (RF) communications require applications which can transmit more and more power. The new generations of power transmitters which use solid state technologies (Gallium Arsenide (GaAs), Gallium Nitride (GaN), Laterally Diffused Metal Oxide Semiconductor (LDMOS)) require high-power transistors. To properly manage parameters such as size, weight and cooling requirements, these components must improve many parameters such as electrical efficiency and power density while maintaining a high level of reliability and miniaturization. Due to

---

Chapter written by Eric JOUBERT, Olivier LATRY, Pascal DHERBECOURT, Maxime FONTAINE, Christian GAUTIER, Hubert POLAERT and Philippe EUDELINE.

these constraints, the behavior of these components that are subject to high thermal stress becomes increasingly difficult to characterize. There is currently a high demand to fully understand the impact of thermal parameters on the overall performance, on the lifetime and ultimately on the reliability. That is why the development of measurement tools that characterize these phenomena is becoming a necessity. The current measurement methods are based on various effects: infrared (IR) [MCN 06], thermal reflectivity [FON 11, DIL 98, FON 08], current–voltage (I(V)) [CAI 92], Raman [AUB 04, SAR 06] and liquid crystals [LEE 04]. All these methods have advantages and disadvantages that adapt to specific contexts.

The aim of this chapter is to present an original method for measuring the internal thermal component without contact. This method is fast and can evolve to a displacement measuring system. The advantage of this type of simultaneity is the possibility of estimating the semiconductor expansion coefficient and especially its evolution during the life of the component. This parameter is important for calculation of the reliability of a component. In this chapter, only the thermal part of the measurement system is presented.

As this type of measurement is new, it is validated by comparison with established methods in the literature. Three methods of measurement are compared in this chapter:

– IR;

– electrical, by using a protection diode internal to the component;

– interferometric.

7.2. Experimental setup

A silicon component used in Global System for Mobile Communications (GSM) and General Packet Radio Service (GPRS) cell phones is used as the demonstrator for all measurements.

It is a system-in package (SIP) combining a passive die and an active die attached by flip chip technology in a HVQFN 40 package (Heat-sink, Very-thin Quad Flat-pack No leads, with 40 input/output pins).

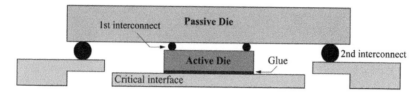

Figure 7.1. *Component used for thermal measurements*

Each input to the active chip of the component used as a demonstrator is protected by an electrostatic discharge (ESD) protection diode. As the characteristic I(V) (current relative to voltage) of a diode depends on temperature, built-in ESD protection diodes can be used as temperature sensors. The component is assembled on a printed circuit board. ESD protection diodes are connected to this circuit. These connections are used either to generate local heating by polarizing the diode or to measure voltage changes that are a function of temperature.

IR and interferometric methods both require access to the surface of the chip. The package of the component under investigation is thus opened using laser opening techniques such as ablation, chemical etching and plasma cleaning. A 500 µm deep and 2 mm square opening is made in the molding compound of the package of the active chip (Figure 7.2). The circuit board on which the component is assembled is placed on a test base plate. The temperature of this base plate is kept constant and controlled by a Peltier module (Figure 7.3).

The ESD protection diode located at pin level 30 (Figure 7.4) is supplied with a current of 200 mA amplitude for 30 s to heat the active chip locally. The temperature is measured simultaneously by the IR measurement technique and the electrical measuring technique.

Figure 7.2. *Component in use*

Figure 7.3. *Experimental setup*

Figure 7.4. *Localization of the protection diode (pin 30)*

7.3. Measurement results

7.3.1. *IR measurements*

IR measurements are made using an IR camera. These IR measurements of the surface temperature are not accurate because they do not take into account the emissivity coefficient of the surface material. Calibration is required to obtain the real component surface temperature. The procedure to achieve this calibration is described in Patent No. EP 08 06074. It consists of:

– making a calibration measurement at a given temperature;

– determining the emissivity of the surface for each pixel;

– making a temperature measurement in operational condition;

– correcting the temperature pixel by pixel using the emissivity data determined in the early steps.

On the raw infrared image (Figure 7.5(a)), the temperature of the hot spot which corresponds to the heat dissipation area of the ESD protection diode is 97°C. After calibration, this temperature is measured at 109°C. In Figure 7.5(b) it can be observed that the tracks of the circuit are visible (light grey tracks around the component). As these tracks are made of copper they conduct the heat and are thus hotter than the circuit (dark grey).

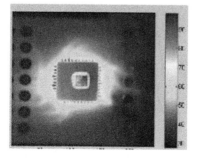 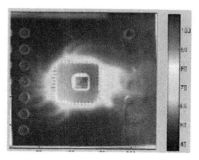

Figure 7.5. *a) Raw IR measure and b) IR measure after calibration. For a color version of this figure, see www.iste.co.uk/elhami/embedded1b.zip*

Calibration measurements are made at four different temperatures: 40, 60, 80 and 90°C. These measurements provide the emissivity at all points on the surface. The results are shown in Figure 7.6.

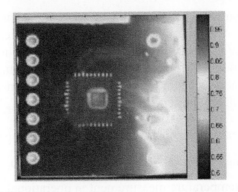

Figure 7.6. *Emissivity matrix of the component. For a color version of this figure, see www.iste.co.uk/elhami/embedded1b.zip*

Since the emissivity is high (0.9) precise measurements can be obtained. The closer the emissivity is to 1 (maximum value equivalent to that of a black body) the more accurate the measurement will be. The graphs in Figure 7.7 show the emissivity of the various zones.

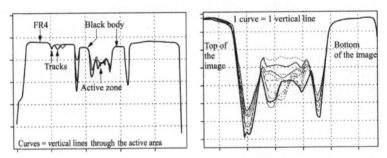

Figure 7.7. *a) Emissivity on a few lines and b) zoom on the active zone*

From these measurements, the average value of the emissivity of the active area is estimated at 0.89 +/–0.02. The results (about one IR picture every 10 s) are shown in Figures 7.8 and 7.9.

Internal Temperature Measurement of Electronic Components 175

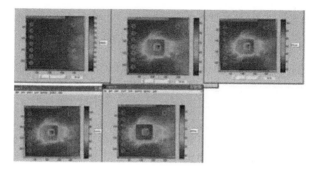

Figure 7.8. *Temperature evolution on the component*

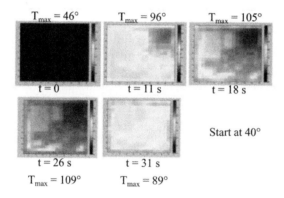

Figure 7.9. *Temperature evolution on the active part*

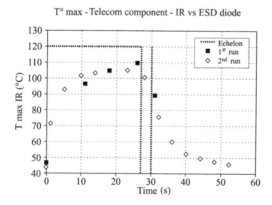

Figure 7.10. *Evolution of temperature against time*

In Figure 7.10, the hot spots of the diode (top-right corner) can be observed (30 s heating time and subsequent cooling).

Several tests were performed. The results are given in Figure 7.11.

7.3.2. Electrical measures

The electrical measurements are made with a SourceMeter Keithley 2601 driven by internally developed software.

7.3.2.1. Diode calibration

The ESD protection diode of pin 30 which is used for temperature measurements is calibrated versus temperature. This calibration is obtained by applying 1 mA current and measuring the voltage at the terminals of this diode at four different temperatures (40, 60, 80 and 90°C) on the test base plate. The calibration results are shown in Figure 7.11(a). The linear regression line obtained from these results giving the temperature (T) as a function of the voltage (V) is: T = 1.2806 × V − 156.43 (Figure 7.11(b)). In this formula, T is expressed in Celsius and V in milliVolt.

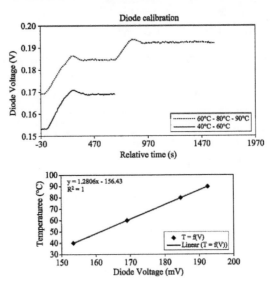

Figure 7.11. a) Temperature calibration results; b) diode voltage versus temperature

7.3.2.2. Temperature measurements using an ESD protection diode

The temperature of the base plate is controlled at 40°C. To heat the chip locally, a 200 mA current is supplied to the protective diode over a 30 s period. Then the chip remains unpowered during a 30 s cooling period. During this period of 60 s, in order to assess the temperature at the diode, the diode voltage is measured every second by applying a current of 1 mA. This allows low current to minimize electric power during the measurement and thus avoid additional heating due to the measurement. A voltage/temperature conversion is then performed by applying the coefficients obtained during the calibration of the diode. The polarization of the diode cycle during the heating period and the cooling period is shown in Figure 7.12.

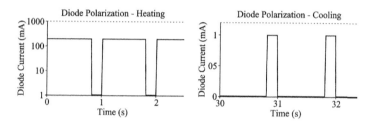

Figure 7.12. *Polarization cycle of the diode during heating and cooling periods*

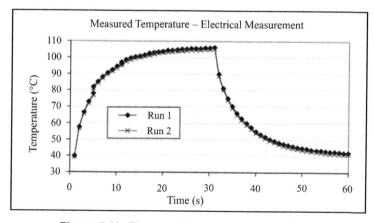

Figure 7.13. *Time evolution of temperature during heating/cooling of the diode*

Two runs of measurements are performed. The results are shown in Figure 7.13. They demonstrate a good reproducibility.

7.3.3. Optical measurement methods

7.3.3.1. Principle

The proposed solution makes it possible to obtain measurements of two quantities: temperature via the refractive index and the reflection coefficient, and expansion coefficient via the phase shift induced on the optical measuring wave through an interferometer system. This simultaneity of measurements is made possible by the use of a Charge Coupled Device (CCD) line sensor. This sensor quantifies the intensity levels over several interference fringes. Software processing can be used to obtain the various developments of the signal.

Microelectronic components, which are based on metals and class III, IV or V materials, have thermal and optical properties that may vary. The measurement methods based on thermal radiation require a surface treatment modifying the emissivity to approach the unit value. The methodology presented enables us to measure several types of semiconductor materials.

The principle of a null incidence device has advantages. By replacing the single-point measurement (fast photodiode) with a line measurement (fast CCD line sensor), more information can be obtained. The Gaussian nature of the beam is easily exploited. The setup is shown in Figure 7.14.

Figure 7.14. *Null optical incidence configuration*

Figure 7.15 shows the divergence of the beam toward the viewing screen. The CCD is positioned according to Figure 7.16.

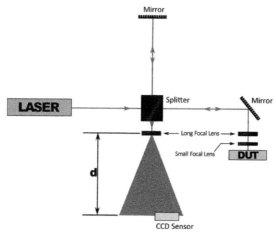

Figure 7.15. *Experimental bench synoptic. For a color version of this figure, see www.iste.co.uk/elhami/embedded1b.zip*

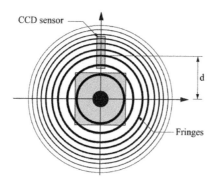

Figure 7.16. *Location of the CCD sensor in the interference pattern*

This interference pattern sees its structure and intensity change according to the following equations:

$$I(x,y) \propto A_1^2(x,y) + R(T).A_2^2(x,y) \\ + 2.A_1(x,y).\sqrt{R(T)}.A_2(x,y).\cos(\phi_2(x,y) - \phi_1(x,y)) \quad [7.1]$$

$$\text{with } R(T) = \left(\frac{n_1 - n_2(T)}{n_1 + n_2(T)} \right)^2 \qquad [7.2]$$

A1 and A2 denote the amplitudes of the electric fields in the two arms of the interferometer, I (x, y) is the intensity received in terms of interference, ϕ_1 and ϕ_2 are the local phase values and R(T) is the optical reflectance coefficient of the component being tested.

These equations show that the interference pattern has a periodic nature. The phase of the signal depends on the distance between the system and the chip being tested, and therefore on the displacement of the component surface caused by temperature. The average amplitude of the interference fringes depends on the reflectance coefficient and thus on the chip surface temperature. Two important parameters are determined independently. Their calculations are validated by simulation (Figure 7.17) and by measurement data (Figure 7.18).

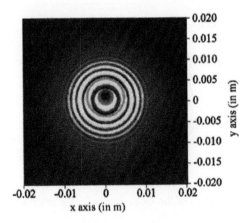

Figure 7.17. *Results obtained from simulation. For a color version of this figure, see www.iste.co.uk/elhami/embedded1b.zip*

Figure 7.18. *Example of a figure obtained from the observation. For a color version of this figure, see www.iste.co.uk/elhami/embedded1b.zip*

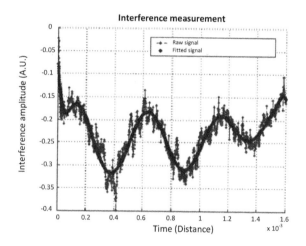

Figure 7.19. *Signal at the output of the CCD sensor and polynomial fitting*

The result measured by the CCD sensor shows a relatively large speckle noise. A polynomial approximation reduces this noise (Figure 7.19).

7.3.3.2. Preliminary results

Preliminary results relate to a standard silicon wafer, placed on a Peltier module that receives a current echelon from t = 0. The Peltier module is fixed on a thermal well which has a temperature rise on its working face (Figure 7.20).

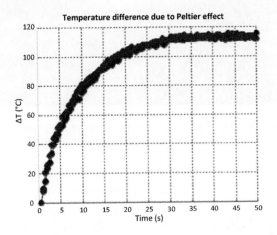

Figure 7.20. *Evolution of temperature on the external face of the Peltier cell*

Analysis of the displacement and amplitude of the interference fringes brings the results in Figure 7.21.

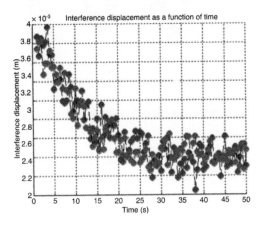

Figure 7.21. *Displacement evolution for a temperature echelon*

Figure 7.22 shows a good homogeneity. The rise in temperature significantly shifts the ripples, and their amplitudes are clearly changed. To be usable, these measures must be calibrated and compared to other types of measurement.

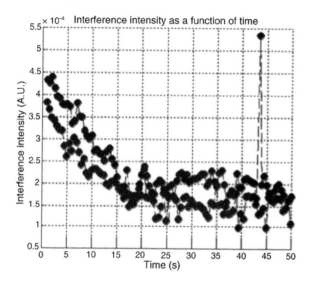

Figure 7.22. *Fringe amplitude evolution for a temperature echelon*

7.3.4. Comparison between infrared and electrical methods

Measuring temperature by the protection diode is carried out for a heating/cooling cycle generated locally on the diode. The whole component is maintained at a temperature of 40°C (temperature of the test base plate). The measurement results obtained by the IR and electrical methods are shown in Figure 7.23.

These measurements show that the results are close. The temperature can be measured with the ESD diodes placed on the component without the need of an IR camera.

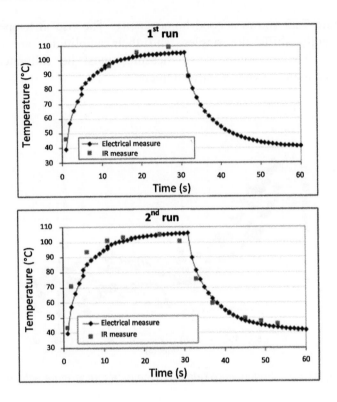

Figure 7.23. *Comparison of electric and IR methods*

It is not possible to use IR temperature and interferometer methods simultaneously. Temperature measurements made using ESD diodes can be used as a reference for another run of measurements carried out using the interferometry method. Figure 7.24 compares temperature results obtained using the ESD diode method and the interferometry method. The observed scatter is due to speckle. This effect can be reduced using spatial filtering. To validate the method, fringe displacement is measured against temperature. Figure 7.25 shows that this displacement varies linearly with temperature. Figure 7.26 shows that the fringe displacement measurement is repeatable. Figure 7.27 shows the thermal model developed for the component under study. Figure 7.28 shows temperature simulation results for this component.

Internal Temperature Measurement of Electronic Components 185

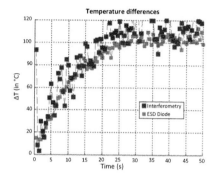

Figure 7.24. *Comparison of ESD/interferometric temperature measurement*

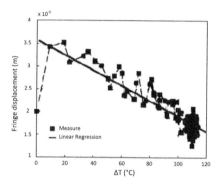

Figure 7.25. *Linearity of displacement measurement*

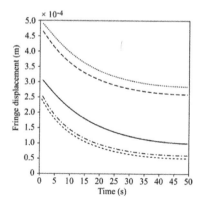

Figure 7.26. *Repeatability of displacement measurement*

Figure 7.27. *Thermal model of the component. For a color version of this figure, see www.iste.co.uk/elhami/embedded1b.zip*

Figure 7.28. *Simulation results for the component. For a color version of this figure, see www.iste.co.uk/elhami/embedded1b.zip*

7.4. Conclusion

The interferometric method presented in this chapter makes it possible to simultaneously measure the physical parameters of an electronic component surface, such as the reflection coefficient and the surface displacement. Using a reverse method based on mathematical models of the interference shapes and fringe measurements, absolute values of the temperature and thermal expansion coefficients can be determined from these observed parameters.

The measurement response time depends on the electronic components and transducers of the setup. Using an ultrafast line CCD sensor or a line of fast photodiodes would make it possible to perform real-time chip temperature measurements. An advantage of this method is that it does not depend on the emissivity of the surfaces found on microelectronic chips. This method is thus more efficient than the infrared-based methods which require powder to be deposited on the surfaces of the chip component under study.

The method based on an ESD protection diode can be used whenever the component under study has such a connection pin. An advantage of this method is the ability measure temperature on the component surface at a location which may be close to the hot spot of the chip. Unfortunately, not all components have such accessible built-in diodes.

Table 7.1 compares the different chip surface temperature measurement methods studied in this chapter.

Characteristics	Thermo IR	Thermo LC	Spectro-Raman	Thermo-reflectometry
Measurement principle	Absolute	Differential	Absolute	Differential
Acquisition time	100 µs	10 ms	>1 min	Real time
Accuracy	1°C	1°C	5°C	0.001°C
Measurement range	500°C	100°C	Very wide	100°C
Measured temperature	Surface	Surface	Internal	Surface
Spatial resolution	5 µm	2 µm	<1 µm	1 µm
2D cartography	Yes	Yes	No	No
Destructive method	No	Yes	No	No

Table 7.1. *Comparison between the different techniques*

7.5. References

[AUB 04] AUBRY R., DUA C., JACQUET J.C., "Temperature measurement by micro-Raman scattering spectroscopy in the active zone of algangan high-electron-mobility transistors", *The European Physical Journal Applied Physics*, vol. 27, pp. 293–296, 2004.

[CAI 92] CAIN B.M., GOUD P.A., ENGLEFIELD C.G., "Electrical measurement of the junction temperature of an RF power transistor", *IEEE Transactions on Instrumentation and Measurement*, vol. 41, pp. 663–665, 1992.

[DIL 98] DILHAIRE S., PHAN T., SCHAUF E. *et al.*, "Sondes laser et méthodologies pour l'analyse thermique à l'échelle micrométrique – application à la microélectronique", *Revue générale de thermique*, vol. 37, pp. 49–59, 1998.

[FON 08] FONTAINE M., JOUBERT E., LATRY O. *et al.*, "New approach for thermal investigation of a III-V power transistor", *Therminic Conference*, Rome, Italy, pp. 26–30, 24–26 September 2008.

[FON 11] FONTAINE M., JOUBERT E., LATRY O. *et al.*, "Simultaneous measures of temperature and expansion on electronic compound", *27th Annual IEEE Conference on Semi-Therm*, San Jose, USA, pp. 203–207, March 2011.

[LEE 04] LEE C.C., PARK J., "Temperature measurement of visible lightemitting diodes using nematic liquid crystal thermography with laser illumination", *IEEE Photonics Technology Letters*, vol. 16, pp. 1706–1708, 2004.

[MCN 06] MCNAMARA D., Temperature measurement theory and practical techniques, Analog Devices, AN-892, 2006.

[SAR 06] SARUA A., JI H., KUBALL M., "Combined infrared and Raman temperature measurements on device structures", *CS MANTECH Conference*, Vancouver, Canada, pp. 179–182, 2006.

8

Reliability Prediction of Embedded Electronic Systems: the FIDES Guide

Reliability prediction calculations usually take place in the first phases of the design process. These calculations are performed to verify that the chosen architecture will achieve reliability goals to identify the critical components for reliability and to evaluate the effect of the operating conditions on the failure rate of the components and, if necessary, to establish the renewal rate necessary for operational condition maintenance. Two reliability prediction handbooks are widely used as standards in the electronics industry: the MIL HDBK 217F and the UTE C 80-810. These manuals which are based on statistical assessments of field returns are no longer updated. The French Ministry of Defense (*Direction Générale pour l'Armement*, DGA) encouraged a consortium of French companies in January 2004 to develop a more precise reliability handbook: the FIDES guide. This manual is updated periodically in order to cover technological evolutions to extend its scope and to include any improvements. The most recent version is the FIDES 2009, which is recognized as the UTE C 80-811 standard. A precise and inclusive definition of the life profile, such as the operating conditions and environmental stresses of systems, has made it unique. Finally, the predictive models developed for the various families of components follow the state of the art as technologies evolve. In this chapter, FIDES is applied to predict the reliability of an embedded automotive mechatronic system.

8.1. Introduction

The observed reliability of a system is expressed by the probability of it operating without failures during the time of observation. Precise

---

Chapter written by Philippe POUGNET, Franck BAYLE, Hichame MAANANE and Pierre Richard DAHOO.

data are only possible to obtain after a sufficient amount of operating time in real usage conditions (field return experience). However, before releasing a mechatronic system on the market, it is wise to have a correct estimate as to its reliability. For systems that cannot be repaired, this reliability assessment evaluates the cost corresponding to the replacement of failed systems during the warranty period. For repairable systems, it estimates the costs and logistics associated with their maintenance in operating conditions.

The failure rate concept is used to predict the reliability of electronic components. The failure rate of a component or a system is defined by the number of failures which occur per unit of time, at a given age. This instantaneous failure rate is expressed in failure in time (FIT) units (1 FIT = 10^{-9} h). Experience shows that for electronic components and systems, the failure rate decreases at the beginning of the lifetime, stabilizes during a long period and increases until the end. The period in which the failure rate is constant corresponds to the useful life period.

Reliability electronic guides are used to calculate the reliability of electronic systems. These predictive reliability calculations propose, for each component, values for their rate of failure, which are adjusted by factors depending on their application and environment, construction and quality, as well as the thermal or electrical stresses they are subject to. The predictive reliability calculation guides widely used in the electronics industry (MIL HDBK 217 and UTE C 80-810) have not been updated for 15 years. However, since the 2000s, the design and technology of electronic component chips has improved and the reliability of the components has increased. The failure rate calculations provided by these guides no longer correspond to field return data. That is why in January 2004 the French Ministry of Defense (*Direction Générale pour l'Armement*, DGA) encouraged the consortium of French companies to propose the FIDES predictive reliability guide. This manual is updated periodically. Today, it is known as the FIDES 2009, UTE C 80-811 standard.

This chapter presents the approach taken to develop the FIDES predictive models and its application to an embedded automotive mechatronic system.

8.2. Presentation of the FIDES guide

8.2.1. *Global modeling*

The FIDES guide takes into account most of the failures encountered in electronic systems in operating conditions, in other words, the failures due to defective components or low-performance batches, as well as those due to electric stresses. However, the reliability prediction calculations of FIDES are accurate if the following conditions are met:

– the failure mechanisms due to aging have no effect either during the early life or during the useful life periods;

– the design of the system is optimized by adopting best practices such as derating, worst-case analysis, highly accelerated testing (HALT), etc.;

– the environmental stress screening (ESS) tests are correctly carried out;

– the operating and environmental conditions of the life profile are specified.

8.2.2. *Generic model*

Each family of components has a failure rate (λ), which is calculated using the following generic formula:

$$\lambda = \lambda_{Physical} \cdot \Pi_{PM} \cdot \Pi_{Process} \qquad [8.1]$$

where $\lambda_{Physical}$ is the failure rate due to the levels of physical stresses the component is subjected to, Π_{PM} represents the quality of the component and the know-how of its manufacturer, $\Pi_{Process}$ describes the impact of the system's lifecycle, in other words its various phases (specification, design, manufacturing, integration and operation).

When manuals such as MIL HDBK 217F were developed, experience had shown that 90% of the failures were due to components and 10% were due to the influence of the system's lifecycle. Recent field return data based on many component failure analyses prove that the situation has evolved. This is due to the component manufacturers' more precise understanding of the physics of failure and to a better control of the manufacturing processes.

Strictly speaking, the metrics of reliability have an additive form (logical OR), whereas the available models are heterogeneous. Failures of a physical origin can be modeled by the fundamental physics laws. Failures due to other causes can only be assessed by conducting audits throughout the lifecycle of the system. A combination of the logical "AND" and "OR" approaches (multiplication and addition) has proven to be the best solution.

8.2.3. *Mathematical foundations*

The FIDES methodology can be applied to both non-repairable and repairable systems. This specificity introduces a difference in the reliability objectives and therefore, in the mathematical models applied.

A non-repairable system is a system which is thrown out after its first failure, since no maintenance is possible. This occurs, for instance, during the assessment of the reliability of a component by accelerated testing or when we consider space applications (civil or military). The reliability objective is then "the probability that the specified mission is successfully achieved". The reliability metric most suited to quantify this objective is the reliability function *R(t)* defined by:

$$R(t) = \exp(-\int_0^t \lambda(u)du) \qquad [8.2]$$

where λ is the sum of the failure rates λ_i corresponding to each independent event.

Repairable systems are kept in operating conditions by maintenance. The reliability objective is then defined as the "mean time between failures" (MTBF) and this applies for the set of systems in operation. The reliability metric most suited for quantifying this objective is the failure intensity.

It must be noted that the failure rate is deterministic, whereas the failure intensity is random, since its trajectory depends on when failures occur and this cannot be known in advance. Because a random character cannot be modeled, therefore, the MTBF parameter cannot be quantified. In order to use a deterministic parameter, the expected value of the number of failures over timer is used instead. If the number of failures is random, its expected value, which is the average number of failures per unit time, denoted by rate of occurrence of failures (ROCOF) is deterministic and modeling is then possible. The reliability measurement for repairable systems is thus assessed from the MTBF parameter expressed as:

$$MTBF = \exp(-m(t)) \int_0^\infty \exp(m(u))du \qquad [8.3]$$

where m(t) is given by:

$$m(t) = \int_0^t \lambda(u)du \qquad [8.4]$$

If λ_i are constant, then the MTBF is expressed as:

$$MTBF = \frac{1}{\lambda} = \frac{1}{\sum_{i=1}^n \lambda_i} \qquad [8.5]$$

8.2.4. *Justifying the use of a constant failure rate/intensity*

Assuming that a constant failure rate/intensity is justified by the following points:

– early failures, which are usually caused by the manufacturing process, are only observed on a small part of components; these failures can be contained by ESS;

– failures due to overstresses or overloads are modeled by an exponential distribution for non-repairable systems and a homogeneous Poisson distribution for repairable systems;

– premature aging due to batches of "defective" or "weak" components is also only observed on a small part of components;

– for the components of repairable systems for which aging is unavoidable, the renewal theory and maintenance activities lead to a ROCOF which is constant in time [RIG 00];

– Drenick's theorem [DRE 60] shows that the constant failure rate/intensity hypothesis is correct when a large number of components and a system in series are considered;

– according to the Occam's razor principle, when the number of failures is small, a simple model (exponential distribution) is more precise than a more complex model (Weibull distribution) [JED 11].

The physical failure rate/intensity ($\lambda_{Physical}$) is based on the Cox model. It is obtained by the following relation:

$$\lambda_{Physical}(X) = \lambda_0 \cdot AF(X) \qquad [8.6]$$

where λ_0 is the failure rate/intensity in reference conditions and AF(X) is the acceleration factor corresponding to the random variable X which expresses the failure factor. The unit of failure rate/intensity is the FIT (1 FIT = 10^{-9} h).

8.2.5. Assessing λ_o

Assessing the failure rate λ_o for every type of component is difficult. The technology of components evolves very rapidly (Moore's law applies since 1964). The physics of failure is in progress as well, and the number of physical failures decreases with time. The failure rate λ_o could be evaluated from specific tests, but this would require huge resources in time and money. The tendency to reduce the lead time to market makes such investment very unlikely. Studies of the reliability of components performed by manufacturers could also be used. However, past experience shows that:

– few failures occur during tests;

– there is a significant variation of the reliability of a component of a given type from one manufacturer to another.

A compromise has been found for the data obtained during the period of manufacturer tests:

– a long enough period of time to be confident as to the assessment of λ_o (a sufficient number of failures);

– an optimized period of time in order to obtain a homogeneous and representative sample of the technology used in the components;

– the largest possible number of manufacturers in order to cover the variability.

8.2.6. Acceleration factors

The acceleration factors are based on the fundamental laws of physics [BAG 02]. The models of these laws are based on general log-linear models (GLLs) [ADA 00]:

$$\lambda(X) = \exp(\alpha_0 + \sum_{i=1}^{s} \alpha_i \cdot g(X_i)) \qquad [8.7]$$

where S is the number of physical stresses and α is a vector of parameters corresponding to the physics of failure laws.

The acceleration factor is then given by:

$$AF = \frac{\lambda(X_0)}{\lambda(X)} = \exp\left[\sum_{i=1}^{s} \alpha_i \cdot (g(X_{0i}) - g(X_i))\right] \qquad [8.8]$$

Table 8.1 gives, for the main physics of failure laws (Arrhenius' law for failures caused by constant temperature levels, the Norris–Landzberg law for thermo-mechanical failures caused by thermal cycles, Peck's law for humidity effects, Basquin's law for vibratory effects and Eyring's law for failures caused by electro-thermal stresses), the correspondence with equation 8.8.

Model	Equation (expression of the function g)	Transformation (expression of the stress X)
Arrhenius (constant temperature)	$\exp\left(\frac{Ea}{KT}\right)$	$X_1 = \frac{1}{T}$
Norris-Landzberg (thermal cycle)	$\exp\left(\frac{Ea}{KT}\right)\left(\frac{a}{\Delta T}\right)^m$	$X_1 = \frac{1}{T}$ $X_2 = Ln(\Delta T)$
Peck (humidity)	$\exp\left(\frac{Ea}{KT}\right)(b \bullet RH)^n$	$X_1 = \frac{1}{T}$ $X_2 = Ln(RH)$
Inverse power law (vibration)	$\left(\frac{a}{G_{res}}\right)^t$	$X_1 = Ln(G_{rms})$
Eyring (electro-thermal)	$\exp\left(\frac{Ea}{KT}\right)(cV)^P$	$X_1 = \frac{1}{T}$ $X_2 = Ln(V)$

Table 8.1. *GLL models for physics of failure laws*

8.2.7. *Life profile*

The life profile describes the type and level of the physical stresses that will be applied on components during their operational life. These data have a direct effect on $\lambda_{Physical}$. In the case of an activity which occurs periodically, a corresponding mean operational activity can be defined. For this reason, the FIDES methodology is applied to one calendar year (8,760 hours) of operational activity.

The physics of failure laws are often expressed relative to a level of physical stress, which is considered independent of time. When the level of stress evolves with time, the life profile is split into phases in which the level of stress is constant. For instance, Figure 8.1 represents the variation of the temperature of the electronics of an embedded automotive mechatronic system as a function of its various modes of use (sleeping mode (parking) and active mode (on) corresponding simultaneously to driving the vehicle and the activation of the system).

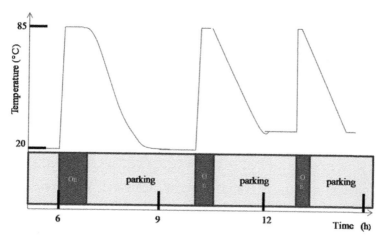

Figure 8.1. *Example of an evolution of the operating temperature*

By breaking down the life profile into phases in which the validity of the physics of failure laws is respected, the level of every physical stress can be set for each of these phases, and the predicted failure rate of the system calculated.

The Sedyakin's principle and the concept of "resource" introduced by Bagdonavicius and M. Nikulin [BAG 02] are used to assess the global failure rate. Following Sedyakin's principle, two levels of physical stress are equivalent if the resources (amounts of life available) being used at time t are equal. Thus, it is possible to find an equivalent temperature in terms of resources, which is independent from time.

Using Arrhenius' law, the failure rate of the system for any given life profile can be assessed. Assuming a constant failure rate, the following equation is obtained:

$$\int_0^\theta \frac{du}{C \bullet \exp\left(\frac{Ea}{KT_{eq}}\right)} = \int_0^\theta \frac{du}{C \bullet \exp\left(\frac{Ea}{KT(u)}\right)} \qquad [8.9]$$

where E_a is the activation energy, K is the Boltzman constant, T is the temperature of the system which varies with time and T_{eq} is the equivalent temperature.

It then follows that:

$$T_{eq} = \frac{-Ea/K}{\ln\left[\frac{1}{\theta}\int_0^\theta \exp\left(-\frac{Ea}{KT(u)}\right) du\right]} \qquad [8.10]$$

Equation [8.10] is valid for any temporal variation of the temperature T(t). Often, the thermal time constant is small with respect to the duration of the phase under consideration. Thus, the previous equation has an analytic solution given the linearity of the "integral" operator. The identified equivalent temperature depends on the life profile and the activation energy.

The failure rate $\lambda_{\text{Physical}}$ of a component depends on the stresses of the life profile and is written as:

$$\lambda_{\text{Physical}} = \left[\sum_{i=1}^{p}\left(\frac{T_{phasei}}{\sum_i T_{phasei}}\right) \bullet \lambda_{phasei}\right] \qquad [8.11]$$

where p is the number of phases in the life profile, T_{phase} is the duration of the phase and λ_{phase} is the failure rate of the phase being considered.

8.2.8. *Testing performed at electronic board level*

FIDES models use physics of failure laws which are derived from accelerated tests performed on electronic boards (Figure 8.2).

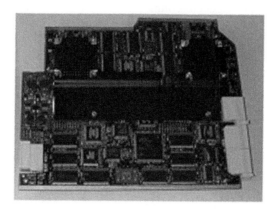

Figure 8.2. *Examples of electronic boards used in the accelerated tests*

For instance, Table 8.2 illustrates the test conditions for characterizing the effect of combined stresses (temperature and humidity). For each level of stress, 16 electronic boards have been tested.

Temperature (°C)	Test N°	Relative humidity (%)
60	1	90
70	2	80
90	3	70

Table 8.2. *Accelerated test conditions*

These test results are modeled using an "Arrhenius – Peck" model (Figure 8.3). The parameters of the "Weibull – Peck" model are evaluated with the maximum likelihood method. The failed components are analyzed to verify that the observed failure mechanisms are the expected ones.

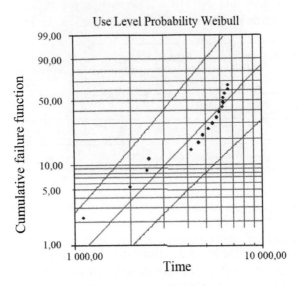

Figure 8.3. *Weibull graph*

The example of discrete active components (Table 8.3) shows that there is a good agreement between the experimental results and the physics of failure laws evaluations, except at certain points. In the case of humidity stress, we observe a significant difference between the activation energies. This discrepancy is due to the conditions of data acquisition, since Peck's law was developed from non-powered electronic boards whereas in these tests, boards were powered. Humidity has little impact because of the self-heating of the components. The failures observed during these accelerated tests are due to both solder joint and connection defects.

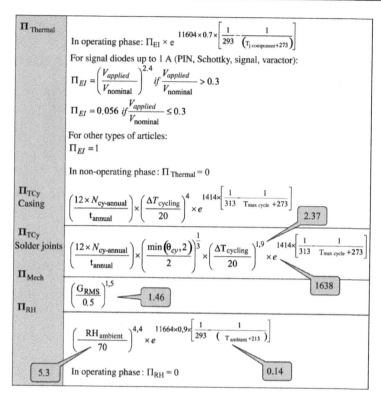

Table 8.3. *Comparison between FIDES parameters and accelerated tests*

8.2.9. Component-level testing

Component-level tests of the FIDES methodology have been carried out on a number of carriers (ground, air and sea). These tests consist of comparing the predicted failure rate with the one observed in operation, relative to the reliability objective, the MTBF or equivalently to the global failure rate of the system.

The assessment is performed by taking the observed data compared to predicted failure rate. A value close to one in the ratio that is obtained indicates that the prediction is accurate. Figure 8.4 shows that the FIDES methodology provides better predictions than MIL HDBK 217F.

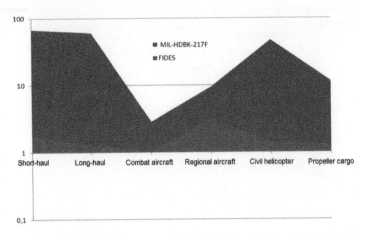

Figure 8.4. *Comparison between system-level predicted and operational reliability. For a color version of this figure, see www.iste.co.uk/elhami/embedded1b.zip*

8.2.10. *"Component family" testing*

"Component family" tests of the FIDES methodology have been carried out on a certain number of carriers (ground, air and sea). These tests consist of comparing the predicted failure rate with the one observed in operation.

The assessment is performed by establishing the ratio of the predicted failure rate with the observed data. A value close to one in the ratio that is obtained indicates that the prediction is accurate. The results are presented by component family, as bar graphs (Figure 8.5). The central bar represents the estimate of the failure rate. The bars on the left-hand side and the right-hand side represent, respectively, the upper and lower bounds of the failure rate for a 90% confidence level. Large differences are observed on the diodes, relays, fuses and piezoelectric devices. For the diodes, a design problem has been identified, generating an abnormally high failure rate.

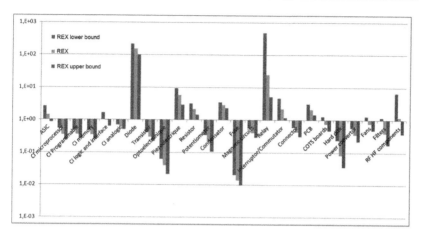

Figure 8.5. *Comparison between predicted and operational reliability by component family. For a color version of this figure, see www.iste.co.uk/elhami/embedded1b.zip*

For the relays, a single failure has been observed, raising considerable uncertainty on the estimate of the failure rate. There are also few relays in a system, in contrast to most of the other components. However, the data (failures and component hours) are insufficient to conclude on the relays. We can make the same kind of statement for the piezoelectric devices.

In the case of the fuses, it is sometimes difficult to determine the cause of failure because of the component's basic function (open circuit in case of overcurrent). For instance, an open fuse may be due to an over current in operation or to a fuse manufacturing defect. When the fuse is not open, it is also difficult to determine whether it performs its required function correctly or not.

8.2.11. *Example of "MOSFET" power transistors*

In order to estimate the referential failure rate λ_o of "MOSFET" power transistors, we use the results of manufacturer tests over an optimal period of time and for a certain number of manufacturers.

8.2.11.1. *Finding the physics of failure law*

The law used to estimate the acceleration factor is based on:

– data about the mechanisms of failure from component manufacturers;

– JEDEC JEP122G standard [JED 11];

– an accelerated test performed on electronic boards.

For active components, an activation energy of 0.7 eV is generally accepted.

8.2.11.2. *Tracking sheet*

Table 8.4 gives a summary of the work carried out from a tracking sheet.

	CONTINUATION SHEET – ACTIVE MODEL		
Active components		SILICON MOS >5W	**Date**: 23/07/2007
Determination of the basic failure rate associated with the chip	HTRB, HTGB tests	Type / sub-family	SILICON MOS > 5W
		Manufacturer test conditions	125°C
		List of manufacturers	OnSemi, ART, NXP, Fairchild, VISHAY
		Observation period	1996 to 2004 & 2006
		Acceleration model	Arrhenius
		Model parameters	Ea = 0.7 eV
		Cumulated hours at manufacturer conditions (h)	9.43E+10
		Number of failures	29
		Basic failure rate at manufacturer conditions	0.392 Fits
		FIDES reference condition	20°C
		Basic failure rate at FIDES conditions	0.020 Fits
Estimated time of maturity of the technology		Technological family	Mature

Table 8.4. *Tracking sheet of the referential failure rate of MOS transistors with power higher than 5 W*

8.3. FIDES calculation on an automotive mechatronic system

The mechatronic system under study is a 2 kW reversible DC-to-DC power converter (DC/DC) dedicated to the operation of a reversible starter-alternator system. Designed to reduce gas consumption this system is located in the engine compartment. It is used in large-inertia vehicles which need supplementary power when starting. The DC/DC converter provides power to the starter-alternator to assist the thermal engine of the vehicle. The power is extracted from a 14 Volt (V) battery and supplies the starter-alternator at a 24 V. This corresponds to the boost operation mode of the converter. When the vehicle is cruising, the 24 V voltage supplied by the starter-alternator is converted into a voltage of 14 V to recharge the battery. This corresponds to the buck operation mode of the converter.

The DC/DC converter is composed of two electronic subsets (Figure 8.6):

– a power board (SMI);

– a control board (FR4).

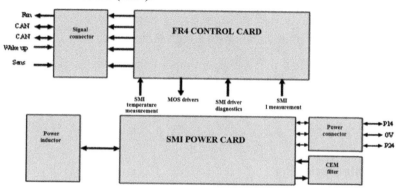

Figure 8.6. *Diagram of the DC/DC converter principle*

The main functions of the SMI board are to:

– boost the voltage of a 14 V network (P14V) to a 24 V network (P24V);

– transfer energy from the 24 V network to the 14 V network.

The main functions of the FR4 board are to:

– control and drive the SMI power board;

– measure the electric current of the SMI board;

– control and monitor metal oxide semiconductor (MOS) drivers;

– measure the temperature of the SMI board;

– receive signals from the connector;

– wake up the convertor when required (start function);

– measure the direction and amplitude of the electric current.

8.3.1. *Goals of the FIDES calculation*

Predictive reliability calculation of the DC/DC converter is performed to:

– assess the predictive failure rate of the system in its operating conditions;

– identify the components with a high failure rate;

– study the effect of various options of the cooling device on the predictive failure rate;

– compare the results obtained by the FIDES manual and the UTE C 80-810 standard.

8.3.2. *Methodology*

Applying the FIDES method consists of the following operations:

– establishing the life profile;

– entering the data for each component of the bill of materials (type, load rate, dissipated power, etc.);

– setting the $\Pi_{manufacturing}$ coefficient (manufacturer quality);

– setting the $\Pi_{process}$ coefficient (quality of the assembly process);

– setting the $\Pi_{induced}$ coefficient which depends on $\Pi_{placement}$, $\Pi_{application}$, $\Pi_{hardening}$ and the sensitivity coefficient.

Once the failure rate of each component is calculated, the global failure rate of the electronic system is obtained by adding the failure rates of each component and the circuits or subsets it is composed of.

8.3.3. *Life profile*

The life profile is defined over a calendar year, in other words over a duration of 8,760 hours. It takes into account the ambient temperatures corresponding to the different seasons, the number of operating cycles and their average durations, the operational stages of the product (sleeping or "parking", active or "on"), the switched off phases, the levels of vibration and chemical pollution and the percentage of relative humidity.

8.3.3.1. *Data entry*

In order to take into account the stresses caused by vibrations, humidity and chemical pollution, the following parameters are entered:

– random mechanical vibration (0.1 G_{rms});

– relative humidity (70%);

– salt pollution (light), artificial pollution (urban area), application area (engine) and protection (hermetic).

The characteristics of the operating stages are then recorded.

8.3.3.1.1. Duration of the phases and number of cycles performed in one calendar year

The specifications of the DC/DC converter define that, in one calendar year, the vehicle is used in driving mode for approximately 2 h. The vehicle is used 335 days/year. In the remaining 30 days, the vehicle is in parking mode.

The durations of the driving and parking phases, as well as the number of cycles performed per day and per year, are given in Table 8.5.

	Per day	Per year
Duration of operation (h)	2	670
Duration of the parking phases	22	8090
Number of days/year		335
Number of cycles	6	2012

Table 8.5. *Duration of the phases (operation and parking) and the number of operating cycles of the DC/DC converter*

8.3.3.1.2. Daily profile

When the vehicle is being used, it performs six trips: one night trip (with a duration of 46 min), one morning trip (with a duration of 25 min) and four daytime trips (each with a duration of 13 min). The DC/DC converter has thus six operating cycles: a cycle of duration 0.77 h, a cycle of duration 0.42 h and four cycles of 0.21 h.

8.3.3.1.3. Distribution of the ambient temperature

In parking mode, the electronic components are switched off. The ambient temperature of the electronics (T_{amb}) is equal to the outside temperature. Over a calendar year, T_{amb} varies according to the seasons. Three seasons can be distinguished: summer, winter and an intermediary season. In order to meet all the climate stresses of the European continent, the external temperature in winter is equal to –10°C, in summer to 30°C and in autumn and spring to 15°C.

In driving mode, the electronics of the converter are switched on and provide power. The thermal engine of the vehicle dissipates heat and causes a rise in the temperature of the engine compartment. The ambient temperature of the converter's electronics (T_{amb}) is equal to the external temperature increased by this rise in temperature.

8.3.3.1.4. Thermal cycles applied on the electronics

In order to evaluate the thermal cycles applied on the electronics, the temperature in the vicinity of the components is determined at the beginning and the end of each operating phase.

In parking mode, the electronic components of the converter are subjected to temperature variation cycles between night- and daytime (10°C).

In driving mode, the amplitude of the thermal cycles ($\Delta T_{cycling\ rate}$) is determined by measuring the temperature increase in the converter casing (T_{casing}) during a power cycle. These measurements are performed on instrumented prototypes providing 2 kW power cycles (Figure 8.7). The average amplitude of the thermal cycles ($\Delta T_{cycling\ rate}$) is equal to 25°C (Figure 8.8).

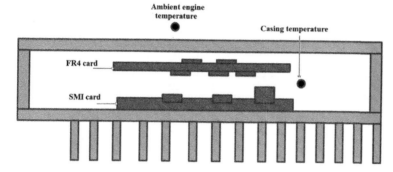

Figure 8.7. *Definition of the ambient engine temperature and the casing temperature on a schematic cross-sectional view of the DC/DC converter*

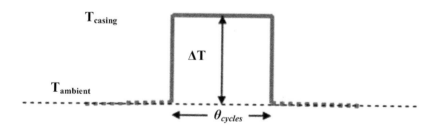

Figure 8.8. *Thermal cycle in driving mode*

The ambient temperature in the vicinity of the components $T_{max\ cycling\ rate}$ is obtained by adding T_{amb} to $\Delta T_{cycling\ rate}$ (Table 8.6). Table 8.7 summarizes the life profile data. It shows, for each phase, the length and number of operating cycles, the amplitude of the thermal cycles and the temperature in the vicinity of the components.

	Winter	Spring–Autumn	Summer
Ambient temperature of the electronics in vehicle sleeping mode (°C)	–10	15	30
Ambient temperature of the electronics in vehicle driving mode (°C)	30	70	85
Local temperature increase due to electronics operation (°C)	25	25	25
Temperature in vicinity of the components (°C)	55	95	110

Table 8.6. *Ambient and maximum temperatures of the DC/DC converter*

Phase		On/off	T_{annual_phase}	$T_{ambient}$	$T_{cycling}$	Length$_{cycles}$	N_{annual_cycles}	$T_{max\_cycling}$	RH	Grms
Garage	Summer	Off	2427	30	10				70	
Garage	Interm. season	Off	4045	15	10				70	
Garage	Winter	Off	1618	10	10				70	0,1
Nighttime	Summer	On	77	85	25	0.77	101	110	70	0,1
Nighttime	Interm. season	On	128	70	25	0.77	168	95	70	0,1
Nighttime	Winter	On	51	30	25	0.77	67	55	70	0,1
Morning	Summer	On	42	85	25	0.42	101	110	70	0,1
Morning	Interm. season	On	70	70	25	0.42	168	95	70	0,1
Morning	Winter	On	28	30	25	0.42	67	55	70	0,1
Daytime	Summer	On	82	85	25	0.21	402	110	70	0,1
Daytime	Interm. season	On	137	70	25	0.21	670	95	70	0,1
Daytime	Winter	On	55	30	25	0.21	268	55	70	0,1

Table 8.7. *Life profile*

8.3.3.1.5. Capturing the components

When the components of the bill of materials are captured, their load rate is set. For resistances and capacitors, the applied voltages and the dissipated powers are set. For active discrete circuits, the junction temperature, the dissipated power and the thermal resistance are set.

8.3.3.1.6. Setting $\Pi_{part\_manufacturing}$

The theoretical range of variation of the $\Pi_{part\_manufacturing}$ factor varies from 0.5 (for the best case) to 2. As components suppliers are integrated into the development process, the $\Pi_{part\_manufacturing}$ factor is set to 0.5.

8.3.3.1.7. Setting $\Pi_{process}$

The $\Pi_{process}$ factor is a representative of the quality and technical control of reliability in the product lifecycle. This factor evaluates the maturity of the industry with respect to its control over the reliability of the engineering process. The theoretical variation range of the $\Pi_{process}$ factor is equal to 1 (for the best process) to 8. Given that the product is mass-produced with a stabilized process, the $\Pi_{process}$ factor is set to 1.

8.3.3.1.8. Setting $\Pi_{induced}$

The $\Pi_{induced}$ factor takes into account the placement of the components on the board, the application conditions and the contribution of accidental mechanical (MOS), electric electrical over stress (EOS) and thermal thermal over stress (TOS) overloads.

Given that the FR4 board is used as a digital interface board and the SMI board is a power board, the placement factor ($\Pi_{placement}$) of the components is set to 1.6. Since the DC/DC converter is located under the engine hood, the $\Pi_{application}$ factor is set to 3. The hardening factor is set to its default value of 1.7. The $\Pi_{induced}$ factor is calculated as a function of the $\Pi_{placement}$, $\Pi_{application}$, $\Pi_{hardening}$ factors and the sensitivity coefficient given in the technical sheet of each component.

8.3.4. Results for the SMI board components

8.3.4.1. Results by component type

The results of the calculations are failure rates expressed in FIT units (1 FIT = 10^{-9} h). These failure rates are given by component type in Table 8.8.

Result by component family	Quantity	Failure rate (FIT)
Capacitor	123	544
Resistance	37	164
Connectors	21	93
Magnetic components	6	27
Active discrete circuits	14	62
Printed circuit board	1	4
Total	202	894

Table 8.8. *Failure rates by component type on the SMI board*

8.3.4.2. Critical components

The most critical components are the power components (chemical capacitors), inductances (self), connections (connection by electric reflow of the Bus Bars), "High Side" and "Low Side" transistors (Buck Boost functions), and X7R-type ceramic capacitors (EMC filtering).

8.3.5. Results for the FR4 board components

8.3.5.1. Results by component type

The results (in FIT) of the predictive reliability calculation of the FR4 board are given in Table 8.9 (1 FIT = 10^{-9} h).

Result by component family	Quantity (Q)	Failure rate (FIT)
Integrated circuits	22	742
Active discrete circuits	136	174
Optocouplers	0	0
Resistance	334	969
Capacitor	156	247
Magnetic components	4	19
Piezoelectric components	1	10
Hermetically sealed electromagnetic relays	0	0
Printed circuit board	1	6
Connectors	0	0
Total	654	2167

Table 8.9. *Failure rates by component type on the FR4 board*

8.3.5.2. *Critical components*

The most critical components are the four integrated circuits and two resistance references.

8.3.5.3. *Connecting wires between the SMI and FR4 boards*

The FIDES standard takes into account the connections between hybrid circuits by wire bonding. The predicted failure rate of the 46 connecting wire bonds is equal to 186 FIT.

8.3.6. *Failure rate of the DC-DC converter*

The predicted failure rate of the DC/DC converter is obtained by making the sum of the failure rates of the SMI and FR4 board components and those of the connecting wires (Table 8.10).

	SMI card	FR4 card	Connecting wires	Converter
Failure rate (FIT)	561	2281	186	3028

Table 8.10. *Failure rate of the converter and its subsets*

8.3.7. *Effect of the amplitude of the thermal cycles on the lifetime*

The amplitude of the thermal cycles ($\Delta T_{\text{cycling rate}}$) depends on the cooling conditions of the power converter casing. Several cooling methods are possible: natural convection or forced ventilation. When the speed of the cooling air forced by ventilation is equal to 2 m/s, the amplitude of the thermal cycles exerted on the electronics is 25°C. In the case of cooling by natural convection, the amplitude of the thermal cycles increases. In order to evaluate the effect of the cooling quality on the predicted failure rate of the converter, we perform predictive reliability calculations by varying $\Delta T_{\text{cycling rate}}$. The choice of the cooling system of the converter (forced or natural convection) has a significant effect on the predicted failure rate of the converter (Figure 8.9).

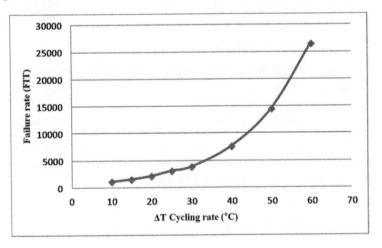

Figure 8.9. *Variation of the predicted failure rate of the DC/DC converter as a function of the ΔT cycling rate*

8.3.8. *Comparison with the results of the UTE C 80-810 standard*

Table 8.11 shows that the failure rates obtained with the UTE C 80-810 standard and the FIDES (UTE C 80-811) are of the same order of magnitude.

	FIDES			UTE C 80-810		
	SMI card	FR4 card	Converter	SMI card	FR4 card	Converter
Failure rate (FIT)	561	2281	3028	954	1798	2752

Table 8.11. *Comparison of predicted reliability calculations between the two standards: FIDES and UTE C 80-810*

8.4. Conclusion

In the electronics industry, two main handbooks are used to evaluate the reliability of electronic systems: MIL HDBK 217F and UTE C 80- 810. However, these two manuals are no longer kept up to date. The FIDES standard which is periodically updated is adapted to the innovative technologies found in mechatronic systems.

In this chapter, the theoretical bases of the FIDES standard and the procedures which have been used to set the values of the failure rates of the components are presented. The FIDES standard is applied to predict the reliability of an automotive mechatronics system, a DC/DC converter. This system is mass-produced and is non-repairable. Its life profile is described over one calendar year in several phases where the level of the applied stress on the components does not depend on time. The thermal and electric loads applied on the components are taken into account. The failure rate estimate provided by the FIDES manual is of the same order of magnitude than the UTE C 80-810 standard.

8.5. References

[ADA 00] ADAMANTIOS M., "Modeling & analysis for multiple stress-type accelerated life data", *Proceedings, Annual Reliability and Maintainability Symposium*, pp. 2–5, 2000.

[BAG 02] BAGDONAVICIUS V., NIKULIN M., *Accelerated Life Tests: Modeling and Statistical Analysis*, Chapman & Hall, Boca Raton, FL, 2002.

[DRE 60] DRENICK R.F., "The failure law of complex equipment", *Journal of the Society for Industrial and Applied Mathematics*, vol. 8, no. 4, pp. 680–689, 1960.

[JED 11] JEDEC STANDARDS, Failure Mechanism and Models for Semiconductor Devices, JEP122G, Arlington, VA, 2011.

[RIG 90] RIGDON E., BASU P.A., "The effect of assuming a homogeneous Poisson process when the true process is power law process", *Journal of Quality Technology*, vol. 22, pp. 111–117, 1990.

[RIG 00] RIGDON E., BASU P.A., *Statistical Methods for the Reliability of Repairable Systems*, John Wiley & Sons, New York, 2000.

9

Multi-objective Optimization in Fluid–Structure Interaction

Fluid–structure interaction (FSI) problems are common and critical in many technical applications, such as wind turbines, aircraft, injection systems or pumps. Optimizing the design is thus of practical importance. Optimization algorithms are used to adjust the parameters of the system under study, taking into account various conditions. This chapter presents a new algorithm for optimizing search retrieval for multi-objective optimization: BSAMO. This evolutionary algorithm (EA) solves real-valued numerical optimization problems. EAs are stochastic research algorithms that are widely used to solve non-linear, non-differentiable complex numerical optimization problems. In order to test its performance, we will apply this algorithm to a well-known multi-objective case study. The FSI will be optimized using a partitioned coupling procedure, and we will test this method on a 3D wing subjected to aerodynamic loads. We will present the Pareto solutions obtained and compare them to those of the Non-dominated Sorting Genetic Algorithm II (NSGA-II). The numerical results demonstrate the efficiency of BSAMO and its ability to solve real-world multi-physics problems.

9.1. Introduction

FSI involves multiple fields in physics and can be found in many engineering systems. FSI impact must be taken into account in design. For example, FSI simulations are carried out to avoid flapping in aircraft and turbomachines [ELH 13, ELH 17], to evaluate the environmental loads and the dynamic response of offshore structures, and to optimize the design of many biomedical applications.

Chapter written by Rabii EL MAANI, Abdelkhalak EL HAMI and Bouchaïb RADI.

The optimization of FSI is of practical importance. Although many studies on numerical methods for FSI problems have been published in recent years, there are few current developments on the optimization of design. The publications presenting this are mainly dedicated to the shape optimization of aeroelastic applications. In [MAU 04], a method for the aerodynamic optimization of a wing profile is developed by manipulating the topology of the structure. In [MAU 03], a partitioned analysis for the calculation of sensitivities is proposed in the context of an optimization problem. Lund *et al.* [LUN 03] optimize the FSI of the Computer Aided Design (CAD) of an academic problem by applying a strong coupling regime.

The task of multi-objective optimization is not to find an optimal solution corresponding to each objective function, but to find a set of solutions called the Pareto optimal method (POM) [COE 06, DEB 14, KON 06]. The use of POM techniques is crucial for real-world applications due to their complexity. The vast majority of real-world applications depend on several variables of a given model. Proper processing of the variables involved is key to a successful optimization.

When the objective function of an optimization problem is non-linear and non-differentiable, evolutionary algorithm (EA) methods are generally used to find the global optimum [KAR 07, ZHA 09]. In an EA, protecting the genetic diversity of a population is very important to the population's ability to support its development in an iterative way. It is generally thought that, in nature, the genetic diversity of a population results from basic genetic processes such as recombination, crossing, mutation, selection and adaptation [DEB 02, KAR 09, NER 10]. Many evolutionary algorithms rely on basic genetic rules such as ABC, the covariance matrix adaptation evolution strategy (CMA-ES) [HUA 16, SMA 18].

Evolutionary algorithms are population-based stochastic research mechanisms that seek near-perfect solutions to a problem. Using a "test individual", an EA tries to turn one individual of the population into another who will fit better. To generate a test individual, EA selects existing individuals as crude genetic material and combines them with various genetic operators. If the test individual has a better aptitude than the original individual, the test individual replaces them in the next generation population.

EAs are effective tools for POMs because they can simultaneously process a set of solutions in a single execution (see, for example, NSGA-II [DEB 02], SPEA-2 [ZIT 01] and NNIA [GON 08]). Recently, a new EA called the backtracking search algorithm (BSA) has been proposed and applied to several real-valued numerical optimization problems [CIV 13]. It uses information obtained from previous generations to search for better shaping solutions. It has been used in the synthesis of concentric circular antenna arrays [GUN 14] and the identification of the parameters of the hyper-chaotic system [LIN 15]. The BSA is attractive because it has a good convergence compared to other algorithms. The Wilcoxon signed-rank test was used to statistically compare the effectiveness of the BSA in solving numerical optimization problems to the performance of six widely used EAs: PSO, CMAES, ABC, JDE, CLPSO and SADE [CIV 13].

In this chapter, we present a BSA which solves multi-objective optimization problems: BSAMO. This algorithm extends BSAs to multi-objective design issues by using two main tools: the non-dominated fast sorting procedure and the crowding distance. The proposed algorithm and its application to FSI can be considered as multidisciplinary design optimization (OMC) with a monolithic architecture. OMC is an area of engineering that focuses on the use of numerical optimization for the design of systems involving multiple disciplines or subsystems [MAR 13].

The chapter is organized as follows: in sections 9.2 and 9.3, we explain some basic concepts of the BSA and multi-objective optimization. Section 9.4 presents the BSAMO and section 9.5 introduces a BSAMO benchmark test for a constrained I-beam with explicit objective functions. Section 9.6 shows the basic equations of IFS with a coupled solution procedure and the IFS optimization process for the application of a 3D aircraft wing subjected to aerodynamic loads. Finally, a conclusion is given in section 9.7.

9.2. Backtracking search algorithm

The BSA is one of the most recently proposed evolutionary algorithms. The motivation for its development was the need for a simpler and more efficient search algorithm with a single control parameter. According to the creator of BSA, the algorithm has good convergence performance compared to other algorithms, is able to solve multimodal problems and is not too sensitive to the initial value of the control parameter [CIV 13].

The performance of the BSA results from a combination of search space exploration and the use of a direction search matrix. The algorithm can be broken down into five successive steps: initialization, selection-I, mutation and crossover operators, and selection-II. These five steps are explained in the following sections.

9.2.1. *Initialization*

To generate an initial population, BSA uses a uniform random distribution function to establish the individuals $\mathbf{P}_{\{i,j\}}$ of the initial population:

$$\mathbf{P}_{i,j} \sim a.(x_{l_j}, x_{u_j}) \text{ for } i=1,\ldots,N \text{ and } j=1,\ldots,D \qquad [9.1]$$

where N is the size of the population, D is the number of decision variables in the problem, x_{lj} and x_{uj} are respectively the lower and upper limits of the *j*-th decision variable and $a := \mathcal{U}(0,1)$ is a random variable that follows a uniform distribution between 0 and 1.

In addition, the BSA requires an archived historical population. It also uses a uniform random distribution for the initial values of the historical population, $\boldsymbol{P}_h$:

$$\mathbf{P}_{h_{i,j}} \sim a.(x_{l_j}, x_{u_j}) \text{ for } i=1,\ldots,N \text{ and } j=1,\ldots,D \qquad [9.2]$$

9.2.2. Selection I

The first selection step uses the historical P_h population to determine the direction of the search. In the same way as for the initial population, the individuals of the historical population are redefined at the beginning of each iteration by the "if-then" rule as follows:

$$\text{if } a < b, \text{ then } P_h := P \mid a, b \sim \mathcal{U}(0,1) \quad [9.3]$$

where $:=$ is the update operation.

Equation [9.3] makes it sure that BSA designates a randomly selected population belonging to a previous generation as a historical population and remembers that historical population until it is modified. Thus, the BSA has a memory. Next, a permutating function is used to randomly modify the order of the individuals in this historical population, as follows:

$$P_h := \text{Permutation}(P_h) \quad [9.4]$$

9.2.3. Mutation operator

The mutation operator generates the initial form of the mutated test population P_m:

$$P_{m_{i,j}} = P_{i,j} + F \cdot (P_{h_{i,j}} - P_{i,j}), \text{ for } i = 1, \ldots, N \text{ and } j = 1, \ldots, D \quad [9.5]$$

where F is a parameter for controlling the magnitude of the search direction matrix, calculated as the difference between the historical and current population matrices $(P_h - P)$.

In this chapter, we use the value $F = \alpha \cdot \mathcal{N}$ where α is a user-defined constant and $\mathcal{N}$ is the standard normal distribution [CIV 13].

9.2.4. Crossover operator

From the test population constructed using the mutation process, the BSA crossover mechanism creates the final form of the test population, denoted as $\mathbf{P}_c$. Based on fitness values assigned by $\mathbf{P}_m$, the best individuals in this population are selected to guide the search among individuals in the target population. The use of the BSA's unique crossing strategy is detailed [CIV 13]:

$$\mathbf{P}_{c_{i,j}} = \begin{cases} \mathbf{P}_{i,j} & \text{if } \mathbf{map}_{i,j} = 1, \\ \mathbf{P}_{m_{i,j}} & \text{if } \mathbf{map}_{i,j} = 0, \end{cases} \quad \text{for } i = 1,\ldots,n_p \text{ and } j = 1,\ldots,D \quad [9.6]$$

where **map** is a binary matrix with integer values $N \times D$ that guides the crossing directions. This matrix determines the individuals of the test population $n\mathbf{P}_m$ to mix with the relevant individuals of $\mathbf{P}_m$.

In addition, the last operation of the crossing step is to repair the individuals of $\mathbf{P}_c$ that can exceed the limits of the allowed search space (that is, $x_{l,j}$, $x_{u,j}$) due to the strategy of BSA mutation. These individuals are regenerated using the BSA limit control mechanism [CIV 13].

9.2.5. Selection II

The BSA performs a second selection step among the test population generated by the crossing step. Thus, individuals in the test population who have better fitness than individuals in the historical population $\mathbf{P}_h$ are used to update the latter. Similarly, if the best individual of $\mathbf{P}$ has a better fitness than the overall minimum value obtained so far by the BSA, the overall minimum is updated to become the best individual and the overall minimum value is updated to become its fitness. The pseudocode of selection II is given in algorithm 1.

```
Algorithm 1: Pseudocode of selection II
Function glob_min = Selection_ II(N, f(x), P_c, P)

1:  P_f := (P | f(x));
2:  P_cf := (P | f(x));
3:  for i = 1 to n_p do
4:      if P_cf(:,i) < P_f(:,i) then
5:          P := P_c;
6:          P_c f(:,i) := P_f(:,i);
7:      end if
8:  end for
9:  P_f_best := min(P_f);  // best ∈ {1,...,N}
10: if P_f best < glob_min then
11:     glob_min := P_f_best;
12: end if
```

9.3. Multi-objective optimization problem

We consider the following multi-objective optimization problem (MOP):

$$\begin{cases} \min_{x \in \Omega} \mathbf{f}(x) = (f_1(x),\ldots,f_m(x))^T \\ \text{under} \quad : \; g_i(x) \leq 0, \; \text{for} \; i=1,\ldots,l \\ \quad\quad\quad\quad h_k(x) = 0, \; \text{for} \; k=1,\ldots,p \end{cases} \quad [9.7]$$

where **f** is the vector of the concurrent objective functions to be optimized, m is the number of objective functions, $x = (x_1,\cdots,x_n) \in \Omega$ is the decision space in D-dimensions where each decision variable x_i is delimited by lower and upper bounds for $x_{li} \leq x_i \leq x_{ui}$ for $i=1,\ldots,D$, g_i (x) are l inequality constraints and h_k (x) are p equality constraints.

In the following sections, we will introduce some basic concepts that will be useful for the rest of this chapter [BOS 03]:

1) *Pareto dominance.* $\mathbf{a} = (a_1,...,a_n)$ and $\mathbf{b} = (b_1,...,b_n)$ are two vectors. Vector **a** dominates vector **b** if and only if **a** is partially inferior to **b**:

$$(\forall k \in \{1,...,m\}: f_k(\mathbf{a}) \leq f_k(\mathbf{b})) \wedge (\exists k \in \{1,...,m\}: f_k(\mathbf{a}) < f_k(\mathbf{b})) \quad [9.8]$$

(denoted $\mathbf{a} \succ \mathbf{b}$).

2) *Pareto-optimal solution.* A solution $\mathbf{a} \in \Omega$ is said to be Pareto-optimal if there exists no solution $\mathbf{b} \in \Omega$ that dominates **a**:

$$\nexists \mathbf{b} \in \Omega : \mathbf{b} \leq \mathbf{a} \quad [9.9]$$

3) *Pareto-optimal set.* The set $\mathbf{P}_S$ of all Pareto-optimal solutions is defined by:

$$\mathbf{P}_S = \{\mathbf{x} \mid \nexists \mathbf{a} \in \Omega : f(\mathbf{a}) \leq f(\mathbf{x})\} \quad [9.10]$$

4) *Pareto front.* The Pareto front, denoted $\mathbf{P}_F$, is the image of the optimal Pareto set $\mathbf{P}_S$ in objective space and is defined as follows:

$$\mathbf{P}_F = \{f(x) : \mathbf{x} \in \mathbf{P}_S\} \quad [9.11]$$

9.4. Proposed algorithm

The BSA was originally developed solely for single-purpose optimization problems. In this chapter, the BSA is extended to make it suitable for handling MOPs. To do this, the main development should concern the fitness values attributed to individuals in the population. This development makes it possible to adapt the main strategies and the simplicity of BSAs to take advantage of the new exploration capabilities of the research space and the use of an adapted search direction matrix, thus increasing efficiency and preserving the spirit of the BSA.

Algorithm 2 gives the proposed pseudo code, where Max$_G$ is the maximum number of generations. This is called BSAMO. It integrates the BSA mutation and crossover operators presented in equations [9.5] and [9.6] and the non-dominated fast sorting and crowding distance from Deb et al. [DEB 02].

The following sections explain the latter element.

Algorithm 2: BSAMO pseudocode

Function P = BSAMO($D, N, m, \mathbf{f}(x), \text{Max}_G, x_l, x_u$)

1: Generate the initial population **P** using 1 and the historical population $\mathbf{P}_h$ using 3;
2: **P** := Find_Non_Dominated_Crowding_Distance ($\mathbf{P}|\mathbf{f}(x)$);
3: **for** $t = 1 : \text{Max}_G$, **do**
4:   [$\mathbf{P}_c, \mathbf{P}_h$] := BSA_Operator($\mathbf{P}, x_l, x_u, \mathbf{P}_h$); // BSA_Operator : Mutation defined in 5 and Crossover in 6;
5:   **P** := Find_Non_Dominated_Crowding_Distance ($[\mathbf{P}; \mathbf{P}_c]|\mathbf{f}(x)$);
6:   Sort and find the current Pareto optimal solution;
7: **end for**

9.4.1. *Fast non-dominated sorting*

The fast non-dominated sorting procedure was developed under the NSGA-II [DEB 02]. In developing this procedure, the dominance count n_p, the number of solutions that dominate the solution p, and the set of solutions S_p that the solution p dominates, are computed for each solution. The first non-dominated front is thus created and initialized with all the solutions having zero as dominance count. Then, for each solution p with $n_p = 0$, each member q of its set S_p is visited and its dominance number is reduced by one. Accordingly, if for any member the dominance count is zero, then it is placed in a separate list Q. The second non-dominated front is then created as a union of all the individuals belonging to Q. The procedure is repeated for the subsequent fronts (F3, F4, etc.) until all individuals have been assigned to their ranks. The fitness is set on a level number; the lower numbers correspond to a higher fitness (F1 is the best).

9.4.2. *Crowding distance*

The crowding distance defined by Deb *et al.* [DEB 02] is used as an estimate of the measure of the diversity of individuals surrounding a given individual (*i*) in the population. This distance is the average distance between two individuals located on either side of the given solution along each objective. Figure 9.1 illustrates the average distance between individuals *i - 1* and *i + 1* at the border of individual *i* located on the Pareto front. This distance is an estimate of the perimeter of the cuboid formed using the nearest neighbors. This metric represents half the perimeter of the cuboid encompassing solution *i*.

The main consideration of the crowding distance is to find the Euclidean distance between each individual on a front according to their *m* objectives. The calculation of the crowding distance, based on the normalized objective values, is given by algorithm 3, where f_m^{max} and f_m^{min} are, respectively, the maximum and minimum values of the *m*-th objective function. The sum of each crowding distance value corresponding to each objective gives the value of the global crowding distance.

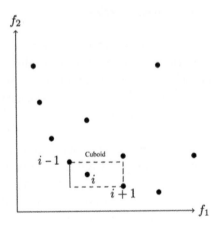

Figure 9.1. *Crowding distance of individual i*

Algorithm 3: Crowding distance calculation for a set solution

1: $n = |\mathcal{I}|$ // number of solutions in $\mathcal{I}$;
2: **for** each i, **do**
3: set $\mathcal{I}[i]_{distance} = 0$;
4: **end for**
5: **for** each objective m, $\mathcal{I} = \text{sort}(\mathcal{I}, m)$ **do**
6: $\mathcal{I}[1]_{distance} = \mathcal{I}[n]_{distance} = \infty$;
7: **for** $i = 2$ to $(n-1)$ **do**
8: $\mathcal{I}[i]_{distance} = \mathcal{I}[i]_{distance} + (\mathcal{I}[i+1]_{.m} - \mathcal{I}[i-1]_{.m}) / (f_m^{max} - f_m^{min})$;
9: **end for**
10: **end for**

In this algorithm, $\mathcal{I}$ is a non-dominated set, n is the number of elements of $\mathcal{I}$, $\mathcal{I}[i]_m$ is the m-th objective value of the individual in $\mathcal{I}$, and the sort$(\mathcal{I}, m)$ is the sorting of individuals $\mathcal{I}$ according to the m-th objective. The theoretical aspect of this algorithm is developed in [ZEI 17].

9.4.3. Numerical validation

9.4.3.1. Benchmark tests

To validate the proposed BSAMO, we used eight multi-objective optimization test functions of well-known references. We used BSAMO to solve these test problems and compared it to the NSGA-II proposed by Deb et al. [DEB 02], which is recognized as a reference in the scientific community of multi-objective optimization. Two metrics are used to compare the performance of the algorithms. We have opted for the generational distance (GD) metric proposed by Van Veldhuizen and Lamont [VAN 98] and the spacing (S) metric proposed by Schott [SCH 95].

– GD: this metric, assuming that $\mathbf{PF}_{true}$ is really available, is a measure representing the "distance" between the approximation front and the true Pareto front. A lower value of GD represents a better performance.

$$GD = \frac{\sqrt{\sum_{i=1}^{n} d_i^2}}{n} \qquad [9.12]$$

where $d_i = \min_j \| PF_{known}(x_i) - PF_{true}(x_j) \|$ refers to the distance in the objective space between the individual x_i and the nearest member in the true Pareto front, and n is the number of individuals in the approximation front.

– S: this metric is a value that measures the regularity of the distribution of non-dominated solutions on the approximation front.

$$S = \sqrt{\frac{1}{n} \sum_{i=1}^{\bar{n}} (d_i - \bar{d})^2} \qquad [9.13]$$

where d_i is the Euclidean distance in the objective space between the individual x_i and the nearest member in the true Pareto front, and $\bar{n}$ is the number of individuals in the approximation front.

Six unconstrained test problems are considered: ZDT1, ZDT2, ZDT3 and ZDT6 (proposed by Zitzler and Thiele [ZIT 98]), Kursawe (proposed by Kursawe [KUR 90]) and Fonseca (proposed by Fonseca and Flemming [FON 98]). Two limited test problems are considered: OSY proposed by OsyczKa and Kundu [OSY 95] and Tanaka proposed by Tanaka [TAN 95]. The Pareto-optimal fronts of these eight problems have several interesting features with convex, non-convex, concave and discontinuous solutions. The population size and the maximum number of generations are both set at 100 for all simulations and both algorithms. We chose the parameter α = 3 for BSAMO. For the NSGA-II, the parameters are 0.9 for crossover probability, 0.5 for mutation probability, 20 for SBX distribution index and 20 for polynomial mutation distribution index. In addition, for comparison purposes, it is ensured that the same initial population is used by both algorithms at each new execution [ZEI 17].

As shown in Figure 9.2, the last non-dominated fronts obtained by BSAMO are more diverse and much better, in terms of generated distance (GD) and S, than those found by NSGA-II for the problems of ZDT1, ZDT2 and ZDT3. For ZDT6, Kursawe and Fonseca, BSAMO obtains a better diversity than NSGA-II. For the Tanaka function, the last Pareto front found by BSAMO has a better GD quality than that obtained by NSGA-II. The Pareto front obtained by BSAMO for the OSY problem is much better than that given by NSGA-II in terms of GD and S. It can be concluded, out of 10,000 function evaluations, that BSAMO performs similar to or better than NSGA-II, GD and S.

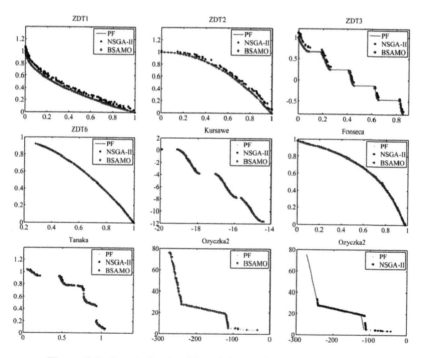

Figure 9.2. *Pareto fronts of the eight test functions. For a color version of this figure, see www.iste.co.uk/elhami/embedded1b.zip*

Table 9.1 provides a statistical analysis of the experimental results for both algorithms and for the eight test problems. It gives the mean, the standard deviation and the median of the GD and S metrics for the results obtained by the two algorithms on 30 independent iterations of the eight test problems.

Problem	Statistic	Generational distance		Spacing metric	
		BSAMO	NSGA-II	BSAMO	NSGA-II
ZDT1	Mean	**1.9332057e-04**	2.9551035e-04	**8.3318911e-03**	9.9576397e-03
	Standard Deviation	5.4992218e-05	4.3787857e-05	1.4874490e-03	6.1334206e-04
	Median	2.0086169e-04	2.9336616e-04	8.3536855e-03	1.0068180e-02
ZDT2	Mean	**9.2557854e-05**	2.2868761e-04	**8.6931978e-03**	1.0378741e-02
	Standard Deviation	6.4796965e-06	3.2438807e-05	1.5567553e-03	7.6266010e-04
	Median	9.2971188e-05	2.3077645e-04	9.2323898e-03	1.0374613e-02
ZDT3	Mean	**1.5686597e-04**	1.8580478e-04	**1.0545011e-02**	2.0146783e-02
	Standard Deviation	1.2808089e-05	1.7098021e-05	8.5796544e-04	7.6755400e-03
	Median	1.5639053e-04	1.8769260e-04	1.0446814e-02	2.0191533e-02
ZDT6	Mean	**7.0749926e-05**	7.1167404e-05	**7.3880130e-03**	8.5770620e-03
	Standard Deviation	4.8712366e-06	3.5178331e-06	9.8349305e-04	3.9006884e-04
	Median	7.0684795e-05	7.1366966e-05	7.6396622e-03	8.5718258e-03
Fonseca	Mean	**2.7989221e-04**	3.0065864e-04	6.5203914e-03	**5.3891371e-03**
	Standard Deviation	3.3520272e-05	3.1116652e-05	6.7429997e-04	4.8175319e-04
	Median	2.7991972e-04	2.9705892e-04	6.5559554e-03	5.4124656e-03
Kursawe	Mean	**1.6745161e-03**	2.1893311e-03	**8.9990255e-02**	8.8457263e-02
	Standard Deviation	2.6183193e-04	3.1584969e-04	1.2727878e-02	1.5233638e-02
	Median	1.6009207e-03	2.1539100e-03	7.7906278e-02	9.1567442e-02
Tanaka	Mean	3.3164734e-01	**3.0189159e-01**	1.4131062e+00	**1.3296550e+00**
	Standard Deviation	8.5356017e-02	4.5854287e-02	1.9489467e-01	2.9204362e-01
	Median	3.1613798e-01	2.9237446e-01	1.3995290e+00	1.2911461e+00
OSY	Mean	**6.7119780e-04**	8.5937344e-04	**6.6138209e-03**	6.7974225e-03
	Standard Deviation	1.0399835e-04	1.1609770e-04	1.4246481e-03	1.1234016e-03
	Median	6.7181790e-04	8.3098530e-04	6.3362175e-03	6.8506880e-03

Table 9.1. *Statistical results of GD and S for the eight test functions on 30 independent iterations*

Since GD is a good measure for evaluating the convergence of an algorithm, these results indicate that BSAMO shows better convergence for seven test problems compared to the NSGA-II.

Moreover, for the Tanaka problem, the GD mean obtained by BSAMO is very close to that obtained by NSGA-II. The results of the S metric show that BSAMO offers the best performance for six problems. For other problems, NSGA-II shows a slightly better performance [ZEI 17].

9.5. Application to FSI problems

9.5.1. *Statement of the FSI problems*

9.5.1.1. *Basic FSI equations*

The FSI problems are presented in a domain Ω composed of a fluid part Ω_f and a solid part Ω_s with respective limits Γ_f and Γ_s and a fluid–structure interface Γ_i. The Eulerian formulation is generally used for the description of the fluid flows, since one is generally interested in the properties of the flows at certain locations in the flow domain. The Navier–Stokes equations of incompressible flows can be written for the fluid domain Ω_f in the form [SOU 13]:

$$\frac{\partial(\rho v)}{\partial t} + \nabla \cdot (\rho v \otimes v - \sigma) - \rho f = 0 \qquad [9.14]$$

$$\nabla \cdot v = 0$$

where ρ is the density of the fluid, v the speed of the external force and σ the stress tensor, defined by:

$$\sigma(v, p) = -pI + 2\mu\epsilon(v) \qquad [9.15]$$

where p is the pressure, I is the identity tensor, μ is the dynamic viscosity and $\varepsilon\,(v)$ is the strain rate tensor given by:

$$\epsilon(v) = \frac{1}{2}\left(\nabla v + \nabla v^T\right) \qquad [9.16]$$

The boundary conditions associated with the fluid domain are:

$$v\big|_{\Gamma_v} = \bar{v} \qquad [9.17]$$

where $\overline{v}$ can be a speed profile known at the limit Γ_v.

Most often, the structure is described using a Lagrangian description, in which the material derivative becomes a partial derivative in time. The structure domain Ω_s is defined by the Navier equation for a homogeneous and isotropic elastic structure like [SOU 13]:

$$\rho_s \frac{\partial^2 u}{\partial t^2} - \nabla \cdot \sigma(u) = 0 \quad \text{in } \Omega_s \qquad [9.18]$$

where ρ_s is the density of the structure, u the displacement and $\sigma(u)$ the stress tensor.

The boundary conditions of the structure are as follows:

$$u\big|_{\Gamma_u} = \overline{u} \qquad \sigma_s\big|_{\Gamma_F} = \overline{F} \qquad [9.19]$$

where Γ_u is the limit where the displacements $\overline{u}$ are imposed and Γ_F is the limit where forces $\overline{F}$ are imposed.

The FSI problem requires the characterization of additional boundary conditions describing the interface between the fluid and the solid. The dynamic and kinematic boundary conditions are the most important criteria:

$$\sigma^f \cdot n = \sigma^s \cdot n \quad \forall x \in \Gamma_i \qquad [9.20]$$

In addition, the normal speeds at the interface must correspond as follows:

$$v \cdot n = \frac{\partial u}{\partial t} \cdot n \quad \forall x \in \Gamma_i \qquad [9.21]$$

9.5.1.2. Fluid–structure coupling

The essential point of a fluid–structure coupled calculation lies in the resolution of the numerical coupling between fluid solver and structure solver, which should be compatible with the modeling of the physical mechanisms responsible for the coupling between the two systems. As such, it is common to distinguish different types of numerical diagrams that induce more or less strong coupling procedures.

The most basic procedure consists of partitioned coupling with alternate exchange between fluid and structure solvers under boundary conditions. This technique has the advantage of being simple from the point of view of its implementation, since the choice of independent solutions for fluids and structures can be free. In a partitioned coupling, the fluid and structure calculations are solved consecutively. It is therefore necessary to provide the displacement of the structure between instants tn and tn + 1 so that the fluid solver can be solved on a domain updated at each time step and vice versa; the fluid charge prediction must be chosen so that the structured solver provides an evaluation of the coherent interface motion with respect to the generated loads [ELM 17, SCH 95].

Conversely, a monolithic coupling procedure makes it possible to simultaneously take into account all the unknowns of the fluid and structure problem in the same instances. In this case, the convergence rate with respect to the coupling is generally optimal [SOBE 13], but the difficulties of these monolithic approaches are of a different order: the difficulty in solving the complete system and the many structural changes that must be made at the level of fluid structure solvers; the loss of software modularity; the limits relating to the application of different sophisticated solvers in different fields; and the challenges related to the size of the problem and the conditioning of the overall system matrix. Therefore, they are generally not considered very well suited to application to real-world problems, where often not only specific solution approaches, but also specific codes, must be used in single fields. For all these reasons, we used a partitioned coupling.

Figure 9.3 gives the iteration process in a diagram view for each time step. After initialization, the flow field is determined in the geometry of the current flow. From this, the friction and pressure forces on the interacting walls are calculated. These are passed to the structure solver as boundary conditions. The structure solver calculates the deformations, with which the fluid mesh is then modified. Then, the stream solver is restarted [SOBE 13].

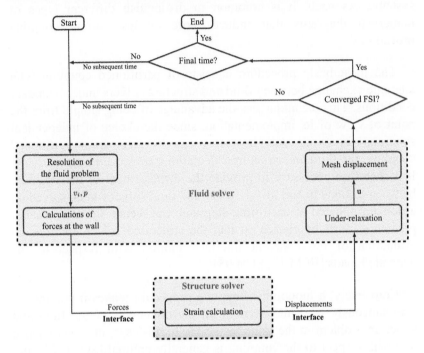

Figure 9.3. *Procedure of fluid–structure coupling*

The FSI iteration loop is repeated until it reaches a convergence criterion ε, defined by the change of the mean displacements:

$$R^{FSI} = \frac{1}{N}\Sigma_{k=1}^{N} \frac{\left\|u_s^{k,m-1} - u_s^{k,m}\right\|_\infty}{\left\|u_s^{k,m}\right\|_\infty} \leq \varepsilon \qquad [9.22]$$

where m is the FSI iteration counter, N is the number of interface nodes and $\|\cdot\|_\infty$ is the infinite norm [SOBE 13].

9.5.2. *Process of FSI optimization*

The flowchart in Figure 9.4 illustrates the process of multi-objective optimization of fluid–structure interaction problems. The optimization work was fully automated through a main script written in MATLAB®. The script starts the BSAMO process by generating an initial population using a uniform random distribution function. Next, the main script submits multiple sets of decision variables (i.e. several geometries) and waits for the values of each objective function before proceeding. MATLAB® calls the ANSYS software to build the volume mesh to calculate the structural responses taken as objective values of the new shape, and returns these values to the optimization process to run the BSAMO algorithm and display the optimal results.

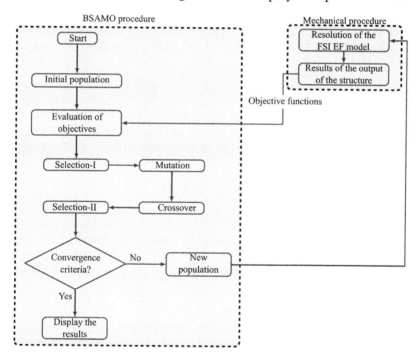

Figure 9.4. *Multi-objective optimization procedure with multi-physics coupling*

9.5.3. Application to the Onera M6 wing

The objective in this case is to minimize the volume and the first eigenfrequency of the wing under a fluid flow subjected to a constraint of its eigenfrequency:

$$\begin{cases} \min_{A_N,\dots} : \begin{pmatrix} \text{Vol} \\ f \end{pmatrix} \\ \text{s.c.} : \ f_1 \leq f_0 \end{cases} \quad [9.23]$$

where f_0 is the upper limit of the maximum allowable natural frequency of the airplane wing.

This formulation allows a controlled determination of the points along P_F by specific adjustment of the values of f_0. A first validation of the proposed algorithm on a plate subjected to a fluid flow appeared in [ELM 19].

Simulation results used in the optimization process are validated by comparing our calculated data with experimental data for the selected wing model. The new computed fluid pressure is transferred to the structure for each new wing shape and iteration of the multi-objective optimization process.

9.5.3.1. Position of the problem

The M6 wing was designed by Bernard Monnerie and his colleagues at ONERA (the French national aerospace research center) in 1972, as part of the cooperation within AGARD (the Advisory Group for Aerospace and Development), to serve as an experimental support for studies of three-dimensional flows at transonic speeds and high Reynolds numbers. The M6 wing was also designed for the validation of numerical fluid mechanics solvers Computational Fluid Dynamics (CFD–conditions representative of the actual flight of military and civilian aircraft). Because of the complexity of transonic flows such as shocks, local supersonic flows and turbulent boundary layer separation, it becomes the most appropriate test for validating CFD solvers. The shape plane of the ONERA M6 wing is shown in Figure 9.5 [SCH 95].

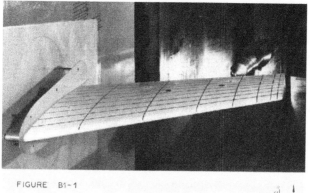
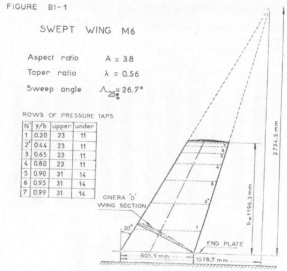

Figure 9.5. *The M6 wing in the ONERA S2MA wind tunnel and its geometric diagram*

Volker Schmitt and François Charpin, scientists at ONERA, recorded the results of these tests in 1979 in an AGARD report. The ONERA-M6 wing results database has been used hundreds of times to validate CFD software and is still used around the world. This is one of the most common way of testing particularly suitable for understanding and evaluating laminar-turbulent transition models, shock-wave boundary layer interaction models, take-off models, etc., which are characteristic phenomena of what happens on the wings near the speed of sound.

9.5.3.2. *FSI numerical modeling*

We have the following data:

– wing surface: we will set this to a so-called boundary condition, setting the speed at 0;

– leading edge: we will adjust this on a symmetrical boundary condition. This basically means that the solution is symmetrical with respect to this plane;

– input, trailing edge, output: we will define the pressure in the far field, the Mach number, the temperature and the speed components. This calculates the speed of the sound and the direction of the speed.

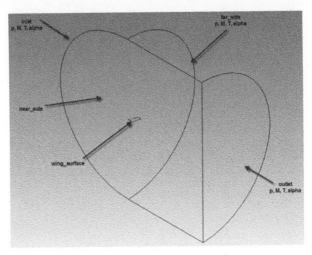

Figure 9.6. *Boundary conditions*

An unstructured hybrid mesh was generated with 375,263 cells and 102,432 volume node points for CFD calculations, as shown in Figure 9.7. It also contains mixtures of tetrahedral, pyramidal and prismatic cells in the boundary layer region. The finite element mesh of the ONERA M6 wing has a total number of 20162 volume node points and 9602 surface node points. The element type for the computational solid dynamics (CSD) mesh is SOLID186 (twenty-node brick element with reduced integration), using only hexahedral elements. The CSD mesh is comparatively rougher than the CFD volume mesh.

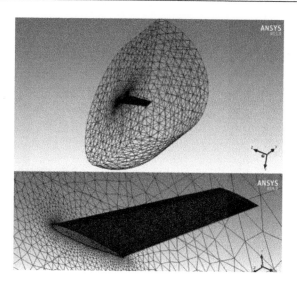

Figure 9.7. *Fluid mesh*

9.5.3.3. *Aerodynamic analysis of the ONERA M6 wing*

Stable aerodynamic analysis was performed at a Mach number (M) of 0.8395 at an angle of attack (α) of 3.06° and a Reynolds number (Re) of 11.72E6. The Spalart–Allmaras (SA) and Menter shear stress transport (MSST) k-ω turbulence models were used for the analysis. These stable aerodynamic calculations were performed to ensure the reliability and accuracy of the internal code for CFD/CSD coupled simulations.

The distribution of pressure on the upper surface of the wing is shown in Figure 9.8. A significant shock was observed on the leading edge near the root and this shock weakened near the tip of the wing, while a strong shock in the middle of the chord was also observed near the wing's trailing edge, which resulted in a form of shock λ. The results of our simulation are validated by comparing the calculated ANSYS / FLUENT data to the experimental data for the ONERA M6 wing [ELM 17, ELM 18].

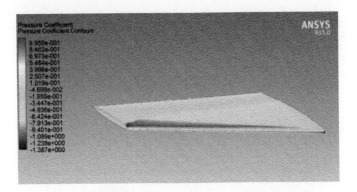

Figure 9.8. *Distribution of pressure on the wing. For a color version of this figure, see www.iste.co.uk/elhami/embedded1b.zip*

In Table 9.2, the drag coefficient and the lift coefficient are compared to those of the NASA WIND simulation [SLA 02] and a good agreement has been observed for a refined mesh. For a refined mesh, the grid has moved from the center of coarse relevance to the center of average relevance, the cell size of the faces of the leading edge and the trailing edge then being refined. The central face of the wing's upper surface and the size of the wing's lower surface have also been refined.

	NASA CFD	**Original mesh**	**Refined mesh**
C_l	0.1410	0.1279	0.134
C_d	0.0088	0.0111	0.0096

Table 9.2. *Comparison of drag (C_d) and lift coefficient (C_l)*

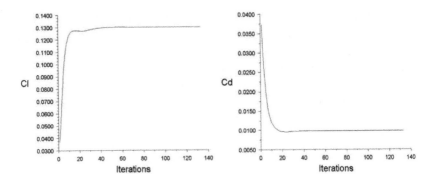

Figure 9.9. *Evolution of drag and lift coefficients*

The pressure coefficient C_p is plotted at different wing locations in the span direction, and the results are compared for the current simulations and the experimental data. Here, the pressure coefficient for locations in the span direction was plotted, $y/b = 0.2$ and 0.9. The resulting C_p values are plotted with experimental data for comparison purposes in Figure 9.10. The experiments were conducted on the M6 wing under transonic flow conditions by [SCH 95].

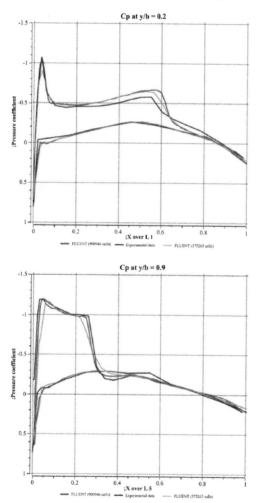

Figure 9.10. C_p for different y/b. For a color version of this figure, see www.iste.co.uk/elhami/embedded1b.zip

It can be seen that in general the shock capture is good and the location of the shockwave is correctly predicted. In the location relative to the scale y / b = 0.2, the shockwave is relatively weaker, and its location and resolution are not as well predicted as in another locations; but as the shockwave becomes steeper along subsequent cross-sections, its location and resolution improve. The overall comparison to the experimental data is good. It can be seen that if we continue to draw the pressure coefficient in the width direction, the results become less accurate due to the three-dimensional effects far from the plane of symmetry. The results of the first coarse mesh and the more refined mesh are compared to the experimental results and a good match is observed [SCH 95].

9.5.3.4. *Aeroelastic analysis of the ONERA M6 wing*

In this section, the deformation due to the aerodynamic load of the wing is considered by performing a unidirectional IFS analysis in steady state [BEN 11]. After developing the aerodynamic load of the wing with ANSYS/FLUENT, the pressures on the appropriate wing areas are transmitted as pressures to ANSYS/Mechanical to determine the stresses and deformations of the wing. The structural configuration considered consists of an aluminum alloy with the following properties: the Young's modulus (E) is 71 gigapascal (GPa), the Poisson's ratio (υ) is 0.32 and the material density is 2770 kg/m^3. The calculated displacement of the deformed and non-deformed wing is presented in Figure 9.11. It is clear that the computed fluid pressure has been successfully transferred to the deformed strain mesh of the CSD.

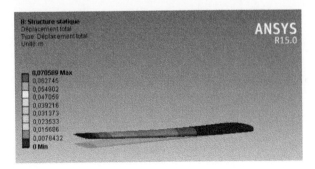

Figure 9.11. *Total displacement of the wing. For a color version of this figure, see www.iste.co.uk/elhami/embedded1b.zip*

9.5.3.5. Pre-stress modal analysis

The problem of eigenvalues and eigenvectors must be solved for mode–frequency analyses. It has the form of:

$$[K]f_i = \lambda_i[M]f_i \qquad [9.24]$$

where [K] is the stiffness matrix of the structure, ϕ_i is the eigenvector, λ_i is the eigenvalue and [M] is the mass matrix of the structure.

For pre-stress modal analyses, the matrix [K] includes the stress stiffness matrix. The results of the modal analysis are shown in Figure 9.12. The result includes the first two modes with their respective eigenvalue values.

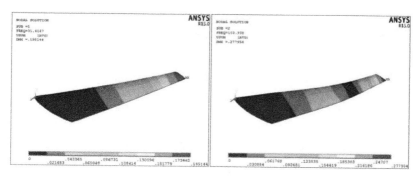

Figure 9.12. *Two first modes of the wing. For a color version of this figure, see www.iste.co.uk/elhami/embedded1b.zip*

9.5.3.6. Optimization

The wing profile set-up is performed with four design variables (A_N, B_N, C_N, D_N), as shown in Table 9.3 and Figure 9.13.

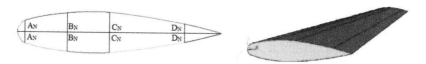

Figure 9.13. *Parameterization of the wing profile*

Design variables	Initial value	Lower bound	Upper bound
A_N	0.054	0.027	0.081
B_N	0.092	0.046	0.138
C_N	0.096	0.048	0.144
D_N	0.044	0.023	0.066

Table 9.3. *Wing design variables*

Figure 9.14 and Table 9.4 show, respectively, the Pareto solutions obtained with the BSAMO and NSGA-II algorithms and the best optimal functions of the ONERA M6 wing, also solved according to the optimization process presented in Figure 9.4. Two of the non-dominated final designs of the aircraft wing are shown in Figure 9.15.

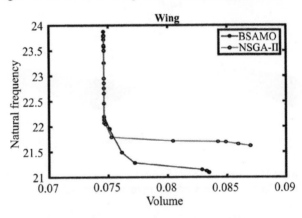

Figure 9.14. *Pareto solutions of the ONERA M6 problem. For a color version of this figure, see www.iste.co.uk/elhami/embedded1b.zip*

		Design variables				Objective functions	
		A_N	B_N	C_N	D_N	Volume	Eigenfrequencies
BSAMO		0.02700000	0.04600000	0.04800000	0.02300000	**0.07462780**	23.72113170
		0.02700000	0.06436623	0.04903415	0.02300000	0.08346423	**21.09539020**
NSGA-II		0.02700342	0.04600303	0.04800069	0.02300822	**0.07463371**	23.79902220
		0.06495546	0.04796975	0.04841059	0.02368821	0.08697976	**21.61767010**

Table 9.4. *Wing's optimal results*

As can be seen, a better quality distribution and an approximate Pareto front are obtained for BSAMO compared to NSGA-II. An examination of the results obtained from the simulations reveals that the success of BSAMO in solving numerical optimization problems is usually not too sensitive to the size or the type of problem. It can be said that BSAMO is statistically more efficient than the other NSGA-II comparison algorithm in solving problems related to explicit objective functions and constraints. In addition, for both multi-physics simulations, it obtains a good quality distribution and approximate Pareto fronts, as well as better optimal results for objective functions.

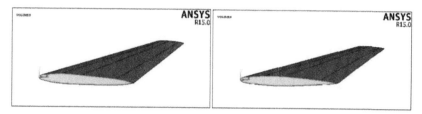

Figure 9.15. *Two of the optimized designs of the aircraft wing*

9.6. Conclusion

In this chapter, an extension of the BSA to solve POMs is presented. The BSA is a simple and efficient global search algorithm with a single control parameter developed for single-objective optimization problems. The proposed multi-objective algorithm, called BSAMO, has the advantage of simply combining BSA operators with non-dominated sorting and crowding distance. Thus, the simplicity obtained preserves the spirit and efficiency of BSA from the combination of exploration of the search space, and use of a direction search matrix.

The success of BSAMO in solving real-world problems was examined in detail by a reference test and then by two multi-physics problems. These consisted of a 2D plate and a 3D airplane wing undergoing FSI phenomena, and treated by coupling the main written script in MATLAB® and the commercial software ANSYS© to calculate the structural responses taken as objective values for the multiple geometries. The results, compared to the NSGA-II algorithm, indicate that BSAMO is generally more efficient and competitive in the processing of complex industrial systems. Therefore, it is concluded that BSAMO offers a potential alternative solution for solving MOPs, as a BSA extension for single-objective optimization issues.

9.7. References

[BEN 11] BENRA F.-K., DOHMEN H.J., PEI J. *et al.*, "A comparison of one-way and two-way coupling methods for numerical analysis of fluid-structure interactions", *Journal of Applied Mathematics*, 2011.

[BOS 03] BOSMAN P., THIERENS D., "The balance between proximity and diversity in multiobjective evolutionary algorithms", *IEEE Transactions on Evolutionary Computation*, no. 2, pp. 174–188, 2003.

[CIV 13] CIVICIOGLU P., "Backtracking search optimization algorithm for numerical optimization problems", *Applied Mathematics and Computation*, no. 15, pp. 8121–8144, 2013.

[COE 06] COELLO C., "Evolutionary multi-objective optimization: A historical view of the field", *IEEE Computational Intelligence Magazine*, no. 1, pp. 28–36, 2006.

[DEB 02] DEB K., PRATAP A., AGARWAL S., MEYARIVAN T., "A fast and elitist multiobjective genetic algorithm: NSGA-II", *IEEE Transactions on Evolutionary Computation*, no. 2, pp. 182–197, 2002.

[DEB 14] DEB K., "Multi-Objective Optimization", in *Search Methodologies*, Springer, New York, 2014.

[ERF 13] ERFANI T., UTYUZHNIKOV S., KOLO B., "A modified directed search domain algorithm for multiobjective engineering and design optimization", *Structural and Multidisciplinary Optimization*, vol. 48, no. 6, pp. 1129–1141, 2013.

[FON 98] FONSECA C.M., FLEMING P. J., "Multiobjective optimization and multiple constraint handling with evolutionary algorithms, part ii: Application example", *IEEE Transactions on Systems, Man, and Cybernetics – Part A: Systems and Humans*, no. 1, pp. 38–47, 1998.

[GON 08] GONG M., JIAO L., DU H. *et al.*, "Multiobjective immune algorithm with nondominated neighbor-based selection", *Evolutionary Computation*, no. 2, pp. 225–255, 2008.

[GUN 14] GUNEY K., DURMUS A., BASBUG S., "Backtracking search optimization algorithm for synthesis of concentric circular antenna arrays", *International Journal of Antennas and Propagation*, 2014.

[ELH 13] EL HAMI A., RADI B., "Comparison study of different reliability-based design optimization approaches", *Advanced Materials Research*, vol. 274, pp. 119–130, 2011.

[ELH 17] EL HAMI A., RADI B., *Fluid–Structure Interactions and Uncertainties: Ansys and Fluent Tools*, ISTE Ltd, London, and John Wiley & Sons, New York, 2017.

[ELM 17a] EL MAANI R., RADI B., EL HAMI A., "Vibratory reliability analysis of an aircraft's wing via fluid–structure interactions", *Aerospace*, no. 3, pp. 1–16, 2017.

[ELM 17b] EL MAANI R., MAKHLOUFI A., RADI B. et al., "RBDO analysis of the aircraft wing based aerodynamic behavior", *Structural Engineering and Mechanics*, no. 4, pp. 441–451, 2017.

[ELM 18] EL MAANI R., MAKHLOUFI A., RADI B. et al., "Reliability-based design optimization with frequency constraints using a new safest point approach", *Engineering Optimization*, vol. 50, no. 10, pp. 1715–1732, 2018.

[ELM 19] EL MAANI R., RADI B., EL HAMI A., "Multiobjective backtracking search algorithm: Application to FSI", *Structural and Multidisciplinary Optimization*, pp. 1–21, 2019.

[HUA 16] HUANG C., RADI B., EL HAMI A., "Uncertainty analysis of deep drawing using surrogate model based probabilistic method", *International Journal of Advanced Manufacturing Technology*, nos 9–12, pp. 3229–3240, 2016.

[IGE 07] IGEL C., HANSEN N., ROTH S., "Covariance matrix adaptation for multi-objective optimization", *Evolutionary Computation*, no. 1, pp. 1–28, 2007.

[KAR 07] KARABOGA D., BASTURK B., "A powerful and efficient algorithm for numerical function optimization: artificial bee colony (ABC) algorithm", *Journal of Global Optimization*, no. 3, pp. 459–471, 2007.

[KAR 09] KARABOGA D., AKAY B., "A comparative study of artificial bee colony algorithm", *Applied Mathematics and Computation*, no. 1, pp. 108–132, 2009.

[KON 06] KONAK A., COIT D.W., SMITH A.E., "Multi-objective optimization using genetic algorithms: A tutorial", *Reliability Engineering & System Safety*, no. 9, pp. 992–1007, 2006.

[KUR 90] KURSAWE F., "A variant of evolution strategies for vector optimization", *International Conference on Parallel Problem Solving from Nature*, pp. 193–197, 1990.

[LIN 15] LIN J., "Oppositional backtracking search optimization algorithm for parameter identification of hyperchaotic systems", *Nonlinear Dynamics*, nos 1–2, pp. 209–219, 2015.

[LUN 03] LUND E., MØLLER H., JAKOBSEN L., "Shape design optimization of stationary fluid–structure interaction problems with large displacements and turbulence", *Structural and Multidisciplinary Optimization*, nos 5–6, pp. 383–392, 2003.

[MAR 13] MARTINS J.R., LAMBE R., ANDREW B., "Multidisciplinary design optimization: A survey of architectures", *AIAA Journal*, no. 9, pp. 2049–2075, 2013.

[MAU 03] MAUTE K., NIKBAY M., FARHAT C., "Sensitivity analysis and design optimization of three-dimensional non-linear aeroelastic systems by the adjoint method", *International Journal for Numerical Methods in Engineering*, no. 6, pp. 911–933, 2003.

[MAU 04] MAUTE K., ALLEN M., "Conceptual design of aeroelastic structures by topology optimization", *Structural and Multidisciplinary Optimization*, nos 1–2, pp. 27–42, 2004.

[NER 10] NERI F., TIRRONEN V., "Recent advances in differential evolution: A survey and experimental analysis", *Artificial Intelligence Review*, nos 1–2, pp. 61–106, 2010.

[OSY 95] OSYCZKA A., KUNDU S., "A new method to solve generalized multicriteria optimization problems using the simple genetic algorithm", *Structural Optimization*, no. 2, pp. 94–99, 1995.

[SCH 95] SCHOTT J.R., Fault tolerant design using single and multicriteria genetic algorithm optimization, Thesis, MIT, 1995.

[SLA 02] SLATER J., ABBOTT J., CAVICCHI R., Validation of WIND for a series of inlet flows, *40th AIAA Aerospace Sciences Meeting & Exhibit*, p. 669, 2002.

[SMA 18] SMAOUI H., ZOUHRI L., KAIDI S., CARLIER E., "Combination of FEM and CMA-ES algorithm for transmissivity identification in aquifer systems", *Hydrological Processes*, vol. 32, pp. 264–277, 2018.

[SOU 13] SOULI M., BENSON D.J., *Arbitrary Lagrangian Eulerian and Fluid–Structure Interaction: Numerical Simulation*, ISTE Ltd, London, and John Wiley & Sons, New York, 2013.

[TAN 95] TANAKA M., WATANABE H., FURUKAWA Y. et al., "GA-based decision support system for multicriteria optimization", *Systems, Man and Cybernetics, Intelligent Systems for the 21st Century, IEEE International Conference*, pp. 1556–1561, 1995.

[VAN 98] VAN VELDHUIZEN, DAVID A., LAMONT G.B., "Multiobjective evolutionary algorithm research: A history and analysis", *Department of Electrical and Computer Engineering Air Force Institute of Technology*, 1998.

[ZEI 17] ZEINE A.T., EL HAMI A., ELLAIA R., PAGNACCO E., "Backtracking search algorithm for multi-objective design optimisation", *International Journal of Mathematical Modelling and Numerical Optimisation*, no. 2, pp. 93–107, 2017.

[ZHA 09] ZHANG J., SANDERSON A., JADE C., "Adaptive differential evolution with optional external archive", *IEEE Transactions on Evolutionary Computation*, no. 5, pp. 945–958, 2009.

[ZIT 98] ZITZLER E., THIELE L., "Multiobjective optimization using evolutionary algorithms – a comparative case study", *International Conference on Parallel Problem Solving from Nature*, pp. 292–301, 1998.

[ZIT 01] ZITZLER E., LAUMANNS M., THIELE L., SPEA2: Improving the strength Pareto evolutionary algorithm, Working paper, TIK-Report 103, Eidgenössische Technische Hochschule Zürich (ETH), Institut für Technische, 2001.

List of Authors

Franck BAYLE
Thales Avionics
CC Navigation
Valence
France

Dan BORZA
LOFIMS
INSA Rouen
Saint-Etienne-du-Rouvray
France

Pierre Richard DAHOO
Université de Versailles Saint-Quentin-en-Yvelines (UVSQ)
Mantes
and
LATMOS – UMR 8190
France

Philippe DESCAMPS
LAMIPS
Caen
France

Pascal DHERBECOURT
GPM
University of Rouen
Saint-Etienne-du-Rouvray
France

Abdelkhalak EL HAMI
LMN
INSA Rouen
Saint-Etienne-du-Rouvray
France

Rabii EL MAANI
ENSAM
Meknes
Morocco

Philippe EUDELINE
Thales Air Systems
Ymare
France

Maxime FONTAINE
GPM
University of Rouen
Saint-Etienne-du-Rouvray
France

Christian GAUTIER
Presto Engineering
Caen
France

Eric JOUBERT
GPM
University of Rouen
Saint-Etienne-du-Rouvray
France

Moncef KADI
IRSEEM
Saint-Etienne-du-Rouvray
France

Alain KAMDEL
LAMIPS
Caen
France

Samh KHEMIRI
IRSEEM
Saint-Etienne-du-Rouvray
France

Malika KHETTAB
LISV
Université de Versailles Saint-Quentin-en-Yvelines (UVSQ)
Vélizy-Villacoublay
France

Ludovic LACHEZE
LAMIPS
Caen
France

Olivier LATRY
GPM
University of Rouen
Saint-Etienne-du-Rouvray
France

Jorge LINARES
GEMAC
Université de Versailles Saint-Quentin-en-Yvelines (UVSQ)
Versailles
France

Hichame MAANANE
Thales Optronique CTG
Elancourt
France

Patrick MARTIN
LAMIPS
Caen
France

Ioana NISTEA
INSA Rouen
Saint-Etienne-du-Rouvray
France

Hubert POLAERT
Thales Air Systems
Ymare
France

Philippe POUGNET
formerly of
Valeo Siemens eAutomotive
Cergy-Pontoise
France

Bouchaïb RADI
LOFIMS
INSA Rouen
Saint-Etienne-du-Rouvray
France

Abhishek RAMANUJAN
IRSEEM
Saint-Etienne-du-Rouvray
France

Zouheir RIAH
IRSEEM
Saint-Etienne-du-Rouvray
France

Index

A, C, D

accelerated test, 4, 19–21, 25
  highly, 19, 20
acceleration factor, 194–196, 204
aerodynamics, 217–219, 239, 242, 247
aeroelastic, 218, 242, 248
aircraft, 217, 219, 236, 244, 245
avalanche, 134, 138, 140–143
calibration, 173, 174, 176, 177, 183
cartography, 66, 68–71, 77–79
cavity, 61
characteristics, 145, 146, 149–151, 153–155, 157, 159–166
characterization, 61
component family, 191, 202, 203, 212, 213
constraint, 124
continuous power wave (CW), 154, 159
controller, 90, 95, 99–101, 103, 105, 106, 108, 109, 120
correlation of digital images (CDI), 16

DC/DC converter, 204–211, 213–215
deformation, 83, 91–100, 102–104, 106, 111, 112, 118, 120
degradation, 123, 126, 128, 133, 138, 139
design optimization, 1
dipole, 61, 72–76, 80, 81
displacement, 83–86, 88, 91–96, 98, 99, 103–105, 108, 112, 113, 115, 119, 120
drain, 146, 147, 149–155, 157–159, 161, 162, 164, 167

E, F, G

electric stress, 159, 164
electromagnetic (EM) stress, 146, 148, 150–152, 159–164
electrostatic discharge (ESD), 123, 124, 126, 127
emissivity, 173, 174, 178, 187
engine control unit (ECU), 6, 7, 9, 11–23, 25, 26
environmental constraints, 29, 31, 53

evolutionary algorithm (EA), 217–220
experimental bench, 179
failure
  mechanism, 1, 4–6, 10–12, 14, 17, 19–26
  mode, 2, 4, 10, 11, 13
  intensity, 193, 194
  rate (λ_o), 189–195, 197, 198, 201–204, 206, 207, 212–215
fluid–structure, 217, 219, 231, 232, 234, 235, 238, 245, 247–249
gain, 146–152, 154, 155, 158, 159, 161, 162, 164
gallium nitride (GaN), 145, 146, 148, 150, 164–167
gap energy (E_g), 54–56
gate voltage, 133, 139
general log-linear (GLL) models, 195, 196
genetic algorithm (GA), 217

H, I, L

HFSS, 68, 72, 76, 77, 81
high electron mobility transistor (HEMT), 145, 146, 148–152, 157, 162, 164, 166, 167
hologram, 83, 84, 86, 87, 90, 93, 101
infrared (IR), 170, 171, 173, 174, 183, 184, 187
interface, 29
interference, 178–182, 187
interferogram, 87, 89, 98, 99, 102–104, 110
life profile, 1, 2, 4, 5, 7, 19, 23, 189, 191, 196–198, 206, 207, 210, 215
linearity, 185

M, N, O

mean time between failures (MTBF), 193, 201
mechatronic module, 37, 56
metal-oxide semiconductor (MOS), 130, 139
  field-effect transistor (MOSFET), 123, 124, 129
microstrip line, 65–68, 70, 76, 77
MIL HDBK 217F, 189, 190, 192, 201, 215
modeling, 2, 4, 5–7, 17, 22–24, 26, 36–45, 47–51, 55, 57
Moiré projection (MP), 84, 112–115
multi-objective optimization, 217–219, 223, 227, 235, 236, 245
multi-physics, 217, 235, 245
non-repairable system, 190, 192, 194, 215
null incidence, 178
optical
  constants, 29, 32, 37, 43, 48–51
  measurement, 178
  optimization, 74–76
  algorithm, 217
overvoltage stress (OVS), 123, 124, 131
  electrical (EOVS), 123, 124, 128

P, R, S

parameters, 126, 133, 135, 137–139
polymer, 29–31, 37, 43, 48, 50–52, 54, 56, 57
power switching, 124
protection diode, 170–173, 176, 177, 183, 184, 187

radiated emission, 61, 65, 72, 81
radio frequency (RF), 145–148,
　151, 154, 156, 158, 160,
　163, 167
　stress, 146, 153–155, 159–161,
　　164
reliability, 1
　matrix, 10, 11
repairable system, 190, 192–194
shielding, 66–68, 70–72, 77–79
　closed, 61, 62, 66–68, 70–72,
　　77–82
　open, 62, 66–68, 70–72, 76–78,
　　80, 81
sintered silver, 29, 43, 45, 56
speckle interferometry technique,
　14, 25
spectroscopic ellipsometry (SE),
　29

static, 146, 149–151, 154, 155,
　157, 159–161, 164
statistical analysis, 230
structured light (SL), 84, 116,
　118, 119

T, U, V, W

thermal, 83, 91, 93–95, 99,
　101–103, 111, 119, 120, 146,
　156, 158
　stress, 145, 156–159
UTE C 80-810 standard, 189,
　190, 206, 214, 215
vibration, 83, 88–92, 94, 95,
　105–109
vibrometry, 16
wrapped phase, 92, 93, 99, 102